Related Books of Interest

Mobile Strategy
How Your Company Can Win by Embracing Mobile Technologies

By Dirk Nicol

ISBN-10: 0-13-309491-X
ISBN-13: 978-0-13-309491-6

Mobile Strategy gives IT leaders the ability to transform their business by offering all the guidance they need to navigate this complex landscape, leverage its opportunities, and protect their investments along the way. IBM's Dirk Nicol clearly explains key trends and issues across the entire mobile project lifecycle. He offers insights critical to evaluating mobile technologies, supporting BYOD, and integrating mobile, cloud, social, and big data. Throughout, you'll find proven best practices based on real-world case studies from his extensive experience with IBM's enterprise customers.

Modern Web Development with IBM WebSphere
Developing, Deploying, and Managing Mobile and Multi-Platform Apps

By Kyle Brown, Roland Barcia, Karl Bishop, Matthew Perrins

ISBN-10: 0-13-306703-3
ISBN-13: 978-0-13-306703-3

This guide presents a coherent strategy for building modern mobile/web applications that are fast, responsive, interactive, reusable, maintainable, extensible, and a pleasure to use.

Using well-crafted examples, the authors introduce best practices for MobileFirst development, helping you create apps that work superbly on mobile devices and add features on conventional browsers. Throughout, you'll learn better ways to deliver Web 2.0 apps with HTML/JavaScript front ends, RESTful Web Services, and persistent data. Proven by IBM and its customers, the approach covered in this book leads to more successful mobile/web applications—and more effective development teams.

Related Books of Interest

Mastering XPages

IBM's Best-Selling Guide to XPages Development—Now Updated and Expanded for Lotus Notes/Domino 9.0.1

By Martin Donnelly, Mark Wallace, Tony McGuckin

ISBN-10: 0-13-337337-1
ISBN-13: 978-0-13-337337-0

Three key members of the IBM XPages team have brought together comprehensive knowledge for delivering outstanding solutions. They have added several hundred pages of new content, including four new chapters. Drawing on their unsurpassed experience, they present new tips, samples, and best practices reflecting the platform's growing maturity. Writing for both XPages newcomers and experts, they cover the entire project lifecycle, including problem debugging, performance optimization, and application scalability.

XPages Portable Command Guide

A Practical Primer for XPages Application Development, Debugging, and Performance

By Martin Donnelly, Maire Kehoe, Tony McGuckin, Dan O'Connor

ISBN-10: 0-13-294305-0
ISBN-13: 978-0-13-294305-5

A perfect portable XPages quick reference for every working developer. Straight from the experts at IBM®, *XPages Portable Command Guide* offers fast access to working code, tested solutions, expert tips, and example-driven best practices. Drawing on their unsurpassed experience as IBM XPages lead developers and customer consultants, the authors explore many lesser known facets of the XPages runtime, illuminating these capabilities with dozens of examples that solve specific XPages development problems. Using their easy-to-adapt code examples, you can develop XPages solutions with outstanding performance, scalability, flexibility, efficiency, reliability, and value.

IBM Press™

Visit ibmpressbooks.com
for all product information

Related Books of Interest

XPages Extension Library
A Step-by-Step Guide to the Next Generation of XPages Components

By Paul Hannan, Declan Sciolla-Lynch, Jeremy Hodge, Paul Withers, Tim Tripcony

ISBN-10: 0-13-290181-1
ISBN-13: 978-0-13-290181-9

XPages Extension Library is the first and only complete guide to Domino development with this library; it's the best manifestation yet of the underlying XPages Extensibility Framework. Complementing the popular *Mastering XPages*, it gives XPages developers complete information for taking full advantage of the new components from IBM.

Combining reference material and practical use cases, the authors offer step-by-step guidance for installing and configuring the XPages Extension Library and using its state-of-the-art applications infrastructure to quickly create rich web applications with outstanding user experiences.

SOA Governance
Achieving and Sustaining Business and IT Agility

Brown, Laird, Gee, Mitra
ISBN: 978-0-13-714746-5

Being Agile
Eleven Breakthrough Techniques to Keep You from "Waterfalling Backward"

Ekas, Will
ISBN: 978-0-13-337562-6

Common Information Models for an Open, Analytical, and Agile World
Chessell, Sivakumar, Wolfson, Hogg, Harishankar
ISBN-10: 978-0-13-336615-0

Disciplined Agile Delivery
A Practitioner's Guide to Agile Software Delivery in the Enterprise

Ambler, Lines
ISBN: 978-0-13-281013-5

Getting Started with Data Science
Making Sense of Data with Analytics

Haider
ISBN: 978-0-13-399102-4

Patterns of Information Management
Chessell, Smith
ISBN: 978-0-13-315550-1

Enterprise Class Mobile Application Development

Enterprise Class Mobile Application Development

A Complete Lifecycle Approach for Producing Mobile Apps

Leigh Williamson   Roland Barcia

Omkar Chandgadkar   Ashish Mathur

Soma Ray   Darrell Schrag

Roger Snook   Jianjun Zhang

IBM Press Pearson plc
New York • Boston • Indianapolis • San Francisco
Toronto • Montreal • London • Munich • Paris • Madrid
Cape Town • Sydney • Tokyo • Singapore • Mexico City
ibmpressbooks.com

IBM Press Program Managers: Steven M. Stansel, Ellice Uffer

Cover design: IBM Corporation

Editor-in-Chief: Dave Dusthimer

Marketing Manager: Stephane Nakib

Acquisitions Editor: Mary Beth Ray

Publicist: Brad Yale

Editorial Assistant: Vanessa Evans

Managing Editor: Kristy Hart

Designer: Alan Clements

Project Manager: Namita Gahtori, Cenveo® Publisher Services

Copy Editor: Cenveo Publisher Services

Indexer: Cenveo Publisher Services

Compositor: Cenveo Publisher Services

Proofreader: Cenveo Publisher Services

Manufacturing Buyer: Dan Uhrig

Published by Pearson plc

Publishing as IBM Press

For information about buying this title in bulk quantities, or for special sales opportunities (which may include electronic versions; custom cover designs; and content particular to your business, training goals, marketing focus, or branding interests), please contact our corporate sales department at corpsales @pearsoned.com or (800) 382-3419.

For government sales inquiries, please contact governmentsales@pearsoned.com.

For questions about sales outside the U.S., please contact international@pearsoned.com.

The following terms are trademarks or registered trademarks of International Business Machines Corporation in the United States, other countries, or both: IBM, the IBM Press logo, developerWorks, IBM MobileFirst, DB2, AIX, WebSphere, Rational, Bluemix, Rational Team Concert, UrbanCode, Worklight, DataPower, Cast Iron, InfoSphere, Optim,

AppScan, IBM z, Tealeaf, CICS, IMS, and Cloudant. Xtify® is a registered trademark of Xtify, an IBM Company. MaaS360 is a registered trademark of Fiberlink Communications Corporation, an IBM Company. A current list of IBM trademarks is available on the web at "copyright and trademark information" as www.ibm.com/legal/copytrade.shtml.

Java and all Java-based trademarks and logos are trademarks or registered trademarks of Oracle and/or its affiliates. Microsoft and Windows are trademarks of Microsoft Corporation in the United States, other countries, or both. Other company, product, or service names may be trademarks or service marks of others.

Library of Congress Control Number: 2015950670

ISBN-13: 978-0-13-347863-1
ISBN-10: 0-13-347863-7

Text printed in the United States on recycled paper at R.R. Donnelley in Crawfordsville, Indiana.

First printing: December 2015

I dedicate the book to Cheryl, my wife, who has always supported every project
that I have undertaken.

—Leigh Williamson

I dedicate the book to my wife Blanca. I love you. Thank you for being
my wife and best friend.

—Roland Barcia

I dedicate the book to my father, Devdatta Chandgadkar, for encouraging
me in taking on new challenges and risks.

—Omkar Chandgadkar

I dedicate the book to the whole family who shares my excitement at being
an author and is proud of me for having completed it. Thanks for all your support.

—Ashish K. Mathur

I dedicate the book to my two sisters Debbi and Pam, who are professional
right-handed writers in their own right, for their "grammar lessons" in my
youth that always inspired me to put my "write" foot forward.

—Roger Snook

I dedicate the book to my wife, Li Xu. She's the ultimate embodiment of faith,
kindness, and patience. I also dedicate it to my kids, Grace, Daniel, and Timothy.
The thought of them keeps me going when the going gets tough.

—Jianjun Zhang

Contents

Preface

About This Book

Every year sees dozens of new information technologies spring into general use. Many of these new types of hardware and software enjoy brief popularity in the IT community, but then fade from view, many times within a few months of their broader industry introduction. However, every decade or so, a new technology comes along with such fundamental impact that it forever changes the IT landscape. Every enterprise, across all industries, becomes motivated to adopt its use. Mobile apps represent one such transformational technology.

Over the course of 35 years in the computing industry, I've participated in four of these transformational IT waves, with mobile computing being the most recent. There are lots of patterns of adoption that are repeated during each wave.

The important adoption pattern for this book is when established enterprises begin the process of integrating the new technology with their existing systems. These companies have made significant investments in previous generations of IT and the new generation has to be knitted with the old instead of starting over "from scratch."

That is why we titled this book *"Enterprise" Mobile App Development*. It has been written specifically to cover issues and topics that arise when mobile app development meets corporate enterprise IT systems. The coauthors have decades of enterprise software-development expertise as well as extensive depth of knowledge in the mobile aspects of their chapter topic. We cover the entire lifecycle involved in enterprise mobile app development, not just one activity such as coding or testing. The chapters are designed to be useful by themselves, but they all fit into a progression that roughly follows the flow of mobile-development activities in a project.

The IBM® developerWorks® Series

This book continues the line of IBM Press publications that comprise the IBM developerWorks series.

We've seen that as new technologies such as mobile, cloud, and social computing technologies have developed, there is a need for books aimed at the enterprise IT professional level that offer practical, hands-on coverage. Thus, we intend to meet that need with topics in the revamped IBM developerWorks series—this book is the second in the series.

Just as IBM developerWorks has always provided the most up-to-the-minute information on topics of interest to developers, we want the books in this series to provide the best combination of in-depth instruction and links to new and updated material on the web so that the books will both inform our readers on the subjects that interest them and help readers follow along with exercises and examples even when the underlying technologies and products change.

So one of the key aspects of the books in this new series is that we not only provide links to information on developerWorks that are relevant to the topics in the text, but we also provide a "landing pad" about each book on developerWorks that links to constantly updated instructions for installing the tools, working through the examples, and helping developers understand what they need to do to be effective with the IBM products that the books are about.

You can find the landing page for this book at: www.ibm.com/developerworks/dwbooks/enterprise-mobile/index.html.

We hope you enjoy reading this book as much as we've enjoyed writing it.

Leigh Williamson, August 2015

How This Book Is Organized

- **Chapter 1, "Mobile: The New Generation of Information Technology,"** introduces the reasons why mobile applications are more than just a compelling new technology for enterprises. They motivate business innovation and transformation. Chapter 1 provides some examples of this, as well as an introductory discussion of some of the challenges and considerations that mobile software brings into the enterprise IT space.

- **Chapter 2, "Mobile Development Lifecycle Overview,"** provides a full discussion of the lifecycle for developing enterprise mobile apps. While there are many development tasks that are same between mobile software and other kinds of software, there are also some parts of the mobile lifecycle that are unique. Chapter 2 starts with a quick refresher about software lifecycle concepts in general and introduces DevOps concepts that are crucial for success in a modern development project. In addition to describing the idealized mobile development lifecycle, we also make a nod to pragmatism and cover techniques for migrating from a not-so-perfect lifecycle toward one more aligned with mobile-development considerations.

- **Chapter 3, "Design Quality Is Crucial, Make the Investment Up-Front,"** is an important new topic in the coverage of modern software development. It seems like an automatic assumption that good design is essential for any mobile app. It's widely said that without design, a mobile app will fail to be an effective system of engagement. But what do we really mean by "design" in the context of mobile app development? And how can a development team apply techniques that will produce "good" design outcomes? Chapter 3 has been written by practitioners of new design thinking, specifically in the context of enterprise mobile software. You'll learn specific exercises and techniques that will benefit the entire development team and result in outstanding user-centered mobile apps that will delight the people who ultimately interact with them. Design practices are important for all kinds of products, but unfortunately, this is an aspect of enterprise software development that has frequently lacked investment. The mobile era is changing all that. Do not overlook this chapter and miss the opportunity to learn "battle-tested" techniques for putting design into your mobile project.

- **Chapter 4, "Mobile Application Development,"** covers coding and building the mobile app, which are the tasks that are at the core of the development activities. The techniques, languages, and architecture used for mobile development are rapidly changing, with important best practices for mobile software emerging at a furious pace. Chapter 4 covers a discussion about the factors to be considered when selecting an approach for coding and building your enterprise mobile app. Regardless of future evolution in mobile software and technology, these factors will remain relevant to any enterprise mobile project. We recommend rereading Chapter 4 before settling on this aspect of each new mobile undertaking. Chapter 4 also touches on the subject of Cloud software, which fits with modern mobile app architecture like hand and glove. You'll see a comprehensive architecture for mobile/cloud software systems that includes all of the considerations important for an enterprise class solution. The chapter covers mobile app deployment concerns as well, since those are a higher priority for enterprise class apps where exiting systems need to be involved.

- **Chapter 5, "Mobile Enterprise—Beyond the Mobile End Point,"** picks up from where Chapter 4 leaves the mobile backend topic and goes deeply into multiple enterprise mobile backend systems, covering how the mobile "front-end" can connect to and integrate with them. Typical protocols and application programming interfaces (APIs) are discussed. Security is covered in depth, since it is a topic with which every enterprise needs to be concerned. Management of mobile devices and software is part of the topics in this chapter too.

- **Chapter 6, "A Comprehensive Approach to Testing of Mobile Applications,"** deals exclusively with mobile software quality considerations. It doesn't take very long to think of many of the ways that mobile apps present challenges for testing and verification. We cover the range of techniques available to address these testing challenges, along with a comparison to help understand when one approach is better than the others. For mobile apps, testing is a continuous activity that goes on even after the app has been placed into production and is in use on real end user's devices. Given the damage to enterprise reputation and monetary results possible from a "bad app," mobile quality assurance is a job that's never "done." Chapter 6 discusses how to apply the technology and products in a process that flows in a constant cycle.

- **Chapter 7, "Best Practices of Mobile DevOps,"** "closes the loop" of the development cycle with an in-depth discussion of DevOps best practices. Picking up from the initial DevOps coverage in Chapter 2, this chapter goes into more detail about putting DevOps into practice in a mobile software project. Follow the guidelines laid out in Chapter 7 and you will be able to crank up the velocity of your development team and accelerate delivery of the mobile app; something that every mobile project aspires to achieve.

- **Chapter 8, "Conclusions and Further Readings,"** provides review and reference material for further research on each of the topics.

Acknowledgments

Thank you to Kyle Brown and Steven Stansel as key advisors and supporters of this book project. Thank you both for your patience, perseverance, encouragement, and very good guidance. Also, many thanks to Mary Beth Ray who, as Executive Editor, stuck with us through the winding pathway that marked the development of this book. And, of course, so much credit is due to the fantastic group of coauthors who brought decades of experience and boundless enthusiasm to project.

–Leigh Williamson

I wish to convey that all glory and honor goes to God the Father and Lord Jesus Christ. Thank you to my wife, Blanca Barcia. My children (Alyssa, Savannah, Amadeus, and Joseph), my parents, mother-in-law, and other family members. Thank you to my spiritual brothers and sisters at Fairview Gospel Church. Thank you to all folks who contributed to this book by providing content or reviewing: John Ponzo, Greg Truty, Todd Kaplinger, Gal Schalor, Chris Mitchell, Gang Chen, Heather Kreger (and others who worked on the Mobile Reference Architecture for Mobile), and the whole IBM MobileFirst® team.

–Roland Barcia

I would like to thank my colleague Leigh Williamson, for his guidance and mentorship, and Soma Ray, for her continued support and honest feedback.

–Omkar Chandgadkar

I would like to acknowledge my mentor, guide, and coauthor Leigh Williamson for providing me an opportunity to be part of this endeavor as well as critical support in developing the material. I would like to also acknowledge the Rational® Test Workbench team for reviewing the content.

–Ashish Mathur

I would like to acknowledge my two coauthors Leigh Williamson and Omkar Chandgadkar for being mentors, inspirations, and a great support system during the creative process.

–Soma Ray

About the Authors

Leigh Williamson is an IBM Distinguished Engineer who has been working in the Austin, Texas lab since 1989, contributing to IBM's major software projects including OS/2, DB2®, AIX®, Java™, WebSphere® Application Server and associated family of products, the Rational brand of software offerings, the MobileFirst line of solutions, and the IBM Cloud products and services. He is currently a member of the IBM Cloud Strategy team, influencing the direction for the IBM Cloud portfolio. You can follow Leigh on twitter @leighawillia. He holds a B.S. degree in Computer Science from Nova University and an M.S. degree in Computer Engineering from University of Texas at Austin.

Roland Barcia is an IBM Distinguished Engineer and CTO for the Mobile IBM Cloud Support and Lab Services. Roland is responsible for technical thought leadership and strategy, practice technical vitality, and technical enablement. He works with many enterprise clients on mobile strategy and implementations. He is the coauthor of four books and has published more than 50 articles and papers on topics such as mobile technologies, Bluemix™, IBM MobileFirst, Java, Ajax, REST, and messaging technologies. He frequently presents at conferences and to customers on various technologies. Roland has spent the past 16 years implementing mobile, API, middleware systems on various platforms, including Sockets, CORBA, Java EE, SOA, REST, web, and mobile platforms. He has a master's degree in computer science from the New Jersey Institute of Technology.

Omkar Chandgadkar is an experience designer with a background in computer engineering and Human Computer Interaction. At IBM, he is involved in conducting strategic research for developer tools and designing for the complex challenges of enterprise customers. Through his work, Omkar strives to design experiences that solve user problems and generate business value.

Ashish Mathur is an IBM Senior Technical Staff Member and Lead Software Architect for IBM Rational functional testing tools and has the mission to build the next-generation mobile and desktop web-testing software. He has been working on automated testing software since 1993 contributing to major IBM and Rational testing software, including Rational Test Workbench, RFT, RQM, RPT, and Rational Test manager. He has been in multiple roles in automated testing including that of a tester, consultant, subject matter expert, and a developer of the tools. He works out of the IBM India Software Labs in Bangalore, India.

Soma Ray is a UX strategist with research and design backgrounds. With educational background in Electronics, Business Administration, and Human Computer Interaction from University of Pune, India and University of Michigan, Ann Arbor, Soma has always strived to make technology more accessible and empathetic for its users. She has worked in the enterprise technology industry and currently works for the IBM Design Studio in Austin Texas.

Darrell Schrag is a 27-year software professional having spent significant time in the DoD/Aerospace and Financial Services industries. Darrell joined Rational Software in 1993 to bring software-development practices and tools to successful customers. Darrell continues to contribute to customer success with IBM after its acquisition of Rational Software. Darrell has spent time at IBM as a Rational services consultant as well as a worldwide mobile and DevOps specialist. Darrell is currently a Cloud Advisor in the IBM Cloud business unit, helping customers find their best path forward with IBM cloud solutions.

Roger Snook brings 25 years of software product innovation and consultative engagements across several industries focused on developer and project productivity to drive good business results—good design is good business! Roger is an IBM Certified Expert IT Specialist, Open Group Master IT Specialist, and an OMG Certified UML Professional in the Washington DC/West Virginia area, and holds a B.S. degree in Computer Science from Rensselaer Polytechnic Institute, Troy, New York. You can find Roger on several social networks or volunteering in his local community youth soccer or faith-based activities.

Jianjun Zhang is a Senior Technical Staff Member in IBM's Systems group, Middleware division. In the past number of years, he worked on exciting projects including helping to integrate Worklight into IBM's MobileFirst portfolio to become the foundation of the mobile strategy, leading a SaaS product development that helps business developers create mobile and web applications for departmental use without having to master coding skills. Lately, he is building cloud services to help business individuals and organizations alike to develop and manage cloud applications, Internet of Things devices, and business insights. He has a Bachelor's degree from Fudan University in China and a Master's degree from Northern Illinois University in the United States.

CHAPTER 1

Mobile: The New Generation of Information Technology

In this chapter we discuss the reasons that businesses want to develop mobile apps, and the impact that mobile apps have on how businesses operate. We also cover the challenges posed in the development of mobile apps, as an underlying motivation for the rest of the book.

Why Businesses Are Adopting Mobile Applications

Across the globe, more people are using mobile devices, which are increasingly user-friendly and intuitive, as their primary means of obtaining information and requesting services over the Internet. In addition, most enterprises realize that the users of their business applications have shifted from traditional personal computers (desktops and laptops) to mobile devices (smart phones and tablets) as a means to access web-based information. This applies whether the intended user for the application is a direct customer, employee, or business partner.

This crucial shift in end user behavior has motivated enterprises to develop mobile channels for their existing business applications, and to plan for new kinds of applications that can exploit the unique characteristics of mobile devices. While there certainly is value in producing a mobile app user interface for an existing business application, the users of mobile applications have come to expect more from their mobile experience.

Driving Business Process Innovation

Compelling and successful mobile apps create an experience that fully engages the end user. So-called "systems of interaction," these apps anticipate the desires of the user and take full advantage of the rich collection of data that the new mobile devices offer. Plus, they motivate changes in the business processes used to support the applications. Systems of interaction encompass both the engagement with the end user and also the context that existing systems-of-record enable. In short, they open up huge new avenues for innovation in business and propel new ways for businesses to interact with their various stakeholders.

The avenue for innovative user engagement is not limited to mobile phones and tablet devices. The term "omni-channel" has been coined to refer to applications that offer end user engagement across a spectrum of devices, from phones to tablets to PCs to kiosks to automobiles, and many more forms of human-technology interaction. Each of these different application end points ("channels") offers unique characteristics that enable valuable interaction with the user under circumstances best suited to that specific channel.

For example, consider the potential experience of an airline customer who checks in with the airline website two days before their flight and confirms their seat assignment and also requests an upgrade. Then less than one day before the flight, this customer uses the airline mobile app on their iPad Mini to check on their seat upgrade and "check-in" virtually for their flight (providing greater assurance to the airline that the customer will actually take the flight). Then, within a few hours of the flight, this same customer uses the airline app on their smart phone from the airport to check the gate assignment and double check the departure time for the flight (and check again the status of their upgrade request). The customer might also check if there is an airline lounge at the departure airport, and also check on their current frequent flyer mileage totals. The relationship established with this customer can continue.

During the flight, this customer may use one of their devices to buy WiFi access to the Internet. He might check the new estimated arrival time for the flight and email a colleague to confirm their plans once they arrive.

The interaction does not stop there. Once this customer arrives at the destination airport, he can use the airline system of interaction to track his luggage, check on ground transportation, and deal with navigation to a meeting. The connection with the customer can go on and on for days. He might need to confirm (or change) their return flight. He might want to confirm his frequent traveler mileage status or he might want to plan completely new trips. The customer might be using any variety of end user devices to connect to the airline IT systems and services at any time, day or night.

This is a system of interaction. It is ongoing, not a single transaction. It requires context to be effective. It is intelligent and learns from feedback from the user. It is an example of the "killer app" of the 21st century.

A Formula for Designing Engaging Systems

IBM has devised a formula for creating compelling new systems of interaction that drive business innovation. Here is the equation:

Mobile apps [smartphones, desktops, vehicles, devices, . . .] + interaction

characteristics [. . .] = drives process innovation [detect, enrich, perceive, act]

The formula starts with mobile first. The mobile app is what interacts with the end user and can be deployed to more than smartphones. Modern "mobile apps" have deployment target devices as diverse as cars, TVs, desktops, body monitors, and many more.

Interaction characteristics are applied to the mobile app. We define the compelling interaction characteristics as:

- **Omni-channel**: Available across many different classes of end point device
- **Context and social aware**: To automatically tune the interaction to the place and relationships involved
- **Connected to systems of record**: Such as existing business services and data sources, as well as cloud-hosted third party services
- **Experience driven**: Able to learn and adapt to specific end user responses
- **Highly instrumented**: So that data for analysis can be obtained
- **Rapidly revised**: So that app developers can continuously improve interaction

These characteristics are applied by the system of interaction to drive business process innovation. The methodology that drives business process innovation must:

- **Detect** opportunities to engage customers and employees
- **Enrich** interaction context with historical data and trends
- **Perceive** via "in-the-now" dynamic interaction context from location, time, social media, and other events, and . . .
- **Act** on the insights gained to enable positive business outcomes.

Hence, the design characteristics for using mobile technologies to drive business innovation are simply: Detect, Enrich, Perceive, and Act.

Let us consider a quick example. We will use an app for hailing taxis to illustrate our formula for systems of interaction.

As a consumer, you download the app and set up a personal profile including preferences and payment information. When you need a cab, you press "Hail Cab" within the app. Your current location is captured, and in a few seconds you are informed that a driver named Mike is heading in your direction. This cab can be available if you confirm within the next minute. Perhaps some social factors can help influence your action here. Have your friends had a good experience using Mike for taxi service? When you confirm, you see Mike's actual location as he approaches along with an estimated time of arrival (ETA). And Mike's cab location is updated even if Mike takes a wrong turn along the way. When he arrives at your location, you get in the taxi cab and payment is established to Mike upon conclusion of the ride.

Let us see how our methodology might apply to this example scenario.

- **Detect**: Customer loads the app on their smart phone
- **Enrich**: Get customer profile, favorite cab, and favorite destination
- **Perceive**: Location of the customer, location of the closest cab, other nearby friends heading to the same destination
- **Act**: Connect the cab driver and the customer through notification and establish secure channel to pay through the app

You can see that this example app meets all of the criteria for a good system of interaction, providing all of the dimensions to deliver compelling value to the user.

Unique Challenges for Development of Engaging Applications

The creation of systems of interaction involves some unique requirements and challenges.

Most of the rest of this book provides an in-depth view of IBM's recommendations for planning, developing, testing, and deploying mobile applications. Some of those recommendations are echoed in the following paragraphs, and you can refer to the later chapters for more details.

Form Factors and User Input Technology

The first and most obvious aspect of mobile applications is that the form factor for display and user interaction is significantly different from what is used by other forms of software. A smaller form factor means that the amount of data displayed to the end user, and layout of that data, needs to be tuned to the "real estate" available on the device. Significantly less data may be displayed on some devices and therefore it must be exactly the "right" data (most relevant to what the user needs at that point in the application). This variety in form factors motivates the "responsive design" approach for application presentation, where the same application takes advantage of the display resources available on the device where it is running.

Another obvious physical difference for mobile applications is that the mechanisms for user input are different. Mobile devices have pioneered the use of nonkeyboard "gestures" (e.g., touch, swipe, and pinch) as an effective and popular method of user input. Gestures must be planned for and supported for a satisfying mobile application user experience. In addition to tactile user input, mobile devices are a natural target for voice based user input. In fact, the traditional keyboard typing form of user input is probably the least effective and least popular mechanism for input to the new systems of interaction.

Besides gathering input directly from the end user, new systems of interaction have the capability to receive input from other sources such as geo-location from the GPS component of the device and image information from the camera typically built into the device. These forms of input make mobile apps more powerful and useful than applications with a more limited array of input possibilities, and they must be considered during mobile application design and development.

Usability and User Interaction Design

There are several reasons why usability and user interaction need greater attention in the design of mobile applications. The difference in form factors and user input methods is one. It is much more difficult and time consuming to plan how to display only the data that is precisely necessary than it is to simply display all possible data and let the user visually sift through it for what they want. The mobile app designer, by contrast, has to consider the screen real estate. When an application needs to present a broader scope of data with multiple layers of detail, it is usually better to use a progressive discovery approach that allows the user to "drill down" into incrementally greater levels of detail focused on fewer specific items.

The rich variety of input methods available on mobile devices is another reason that early design work must identify and leverage more efficient ways for input data to be delivered than the simple "just type it in a form" design, which is a default for traditional web and PC applications.

Designers must avoid extensive keyboard typing for mobile apps in order to reduce end user frustration (with drastically smaller touch keyboards and lack of traditional typing feedback). Yet, identifying nonkeyboard ways in which information can be gathered and delivered to the mobile app is a significant design challenge.

In addition, there is still a more subtle reason for paying extra attention to the mobile app design effort. The way in which end users interact with mobile devices and the applications running on them is different from how they interact with stationary PCs (and even laptops). Mobile device users typically hold the device in their hand while also interacting with the immediate circumstances of their physical situation. Mobile users typically cannot concentrate on the mobile app for very long before switching attention to their physical surroundings. The interaction model for users of mobile apps is short, interrupted, and "bursty" (meaning that they need to complete the application task very quickly before switching attention).

All of these factors drive the need for applying user-centered design very early in the mobile app development project. Ideally these usability and design considerations should be codified in the requirements for the mobile application, and then linked to the later stage development deliverables, along with the tests that validate that the user interaction and "consume-ability" of the app is as satisfying as possible.

Choice of Implementation Technology

There is a spectrum of implementation choices for mobile applications on the market, and no one answer is perfect for all situations—each choice has its advantages and disadvantages. So the challenge for mobile development teams is to understand the trade-offs between the technologies, and make a choice based on the specific application requirements.

The choice of implementation technology for a mobile project impacts other decisions related to the application's development, including:

- Limiting choices for development tools
- Team roles and structure
- How the application is tested and verified
- How the app is distributed and delivered to the end user

So, the choice of implementation approach for a mobile application is crucial, and this early stage decision needs to be made very carefully.

Native Application Implementation

A "native" implementation means you are writing the application using the programming language and programmatic interfaces exposed by the mobile operating system of a specific type of device. For instance, a native implementation for an iPhone will be written using the Objective-C language (or more recently the Swift programming language) and the iOS operating system APIs that Apple supplies and supports.

Native application implementation has the advantage of offering the highest fidelity with the mobile device. Because the APIs used are at a low level and are specific to the device for which the application is dedicated, the application can take full advantage of every feature and service exposed by that device.

However, native mobile app implementations are completely nonportable to any other mobile operating system—for example, a native Apple iOS app must be totally rewritten if it is to run on an Android device. That makes this native implementation a very costly way of producing a mobile business application.

Web Applications

Newer smart phones and tablets come with advanced web browsers preinstalled, and it is relatively easy to implement a standard web application with special style sheets to accommodate the mobile form factor and approximate the mobile device "look and feel." Mobile applications implemented using this approach support the widest variety of mobile devices, since web browser support for JavaScript and HTML5 is fairly consistent. There are several commercial and open source libraries of Web 2.0 widgets that help with this approach. In addition, the web programming model for mobile application implementation has an advantage for enterprises that already have developers trained in the languages and techniques for web application development.

The disadvantage of pure web application implementation is that such apps have no access to functions/features that run directly on the mobile device, such as the camera, contact list, and so forth.

Hybrid Mobile Application Implementation

Hybrid mobile application implementation is a compromise between pure native implementation and pure web implementation. You write the mobile apps using industry standard web programming languages and techniques such as HTML5 and JavaScript, but you package the app into a natively installable format that is distributed via the app store mechanism.

Hybrid apps are linked with additional native libraries that allow the app to have access to native device features from the single application code base. Because the bulk of a hybrid application is implemented using device-agnostic technology, most of the code for the application is portable and reusable across many different mobile operating systems. However, small segments of native code can also be integrated with the hybrid app, which means that the developer can decide how much of the app implementation shares a common code base (using the web technology) and how much is device-specific customization (written in native code).

Mobile Application Build and Delivery

Because businesses want to deliver mobile applications into the market quickly, mobile development projects typically have extremely aggressive time lines. Inception-to-delivery time frames of a few months are common. The pressure to deliver mobile apps quickly results in the adoption of agile development methods for most mobile projects.

An important element in agile development practices is continuous integration and builds. Application changes that are delivered by developers need to be processed immediately for all of the mobile operating systems on which the application is required to execute. If the mobile application is a hybrid or native implementation, several different builds of the application need to be triggered each time a change set for the application is delivered by a developer. The build setup and configuration for each supported mobile environment will be different from the others, and it is most likely that a small "farm" of build servers will need to be provisioned and available to handle these builds of the mobile application for multiple operating systems.

Testing

Testing poses another huge challenge for mobile application development, because it represents a step-jump in complexity and cost over more traditional applications. Unlike traditional PC and Web applications, the range of potentially supported mobile devices and release levels is staggering. Test matrices for mobile projects commonly contain hundreds and even thousands of permutations of device, mobile OS level, network carrier, locale, and device orientation combinations.

There are more variables for mobile testing that are not relevant for other kinds of software. The same device model may function in a subtly different way when connected to a different carrier network. And the quality of the network connection can have profound impact on the behavior of a mobile application. Even the movement of the mobile device itself may be an important factor in the behavior of the application (some applications specifically exploit device movement).

The majority of mobile apps are based on multi-tier architecture, with the code running on the device itself serving as the "frontend" client to data, and the services supplied by more traditional middle-tier and data center representing the "backend." Effective and comprehensive testing of mobile apps requires that all tiers of the application be addressed, not only the code on the mobile device. The setup and availability of test versions of the middle tier and backend services can present very large cost and complexity challenges for the testing of mobile applications.

Many mobile projects start by using manual testing approaches, which is the quickest way to begin testing. But you have to buy all of the various mobile devices that you plan to support with the app, and pay someone (likely a team of people) to tediously go through a written script of instructions describing the tests on each of those devices for every build of the application. While manual testing serves an important purpose in providing crucial usability feedback for the app, it is extremely expensive and inefficient.

As an alternative, there are mobile app testing solutions that rely on running an agent program on the device for interaction within an automated execution. This approach has the flexibility of using either real physical devices or emulators for testing, with the added efficiency of automation. However, the test team bears the costs of setting up the devices to be tested and installing the test agent on them.

Enterprise Mobile Development

The brief coverage in the previous section of a few of the challenges faced when developing mobile apps should begin to raise your awareness that there are plenty of new and different concerns to address for mobile, compared to more established kinds of enterprise software. All by itself, the development of code that is to be installed and run on mobile devices poses substantial obstacles. But when that mobile app needs to be a part of a mission critical enterprise IT system, or at least integrate with existing enterprise data and resources, that is an order of magnitude greater level of complexity and headaches.

This book has been written specifically to cover issues and topics that arise when mobile app development meets corporate enterprise IT systems. The coauthors have decades of enterprise software development expertise as well as extensive depth of knowledge in the mobile aspects of their chapter topic. We cover the entire lifecycle involved in enterprise mobile app development, not just one activity such as coding or testing. The chapters are designed to be useful by themselves, but they all fit into a progression that roughly follows the flow of mobile development activities in a project.

Summary

Our aim in writing the book is to combine the concerns that we have learned over years of corporate software development with the fresh and rapid aspects of new mobile app production. Since the mobile software domain is still fast moving and evolving, we have worked hard to emphasize techniques and considerations in this book that will remain constant regardless of the latest technology of the day.

One of the guaranteed elements in the software industry is change. We learn to deal with it by identifying the concepts and concerns that endure at a level above the detailed technical aspects. Mobile apps represent technological disruption in a very big way, especially for enterprise software and those of us devoted to producing it. The authors hope that you finish this book with a new list of enduring ideas for how to make your enterprise mobile app successful and delightful!

CHAPTER 2

Mobile Development Lifecycle Overview

"Without risk there can be no progress" –*George M. Low*

Our collective experiences on projects tell us that we will be learning new things and doing things in new ways as our mobile projects evolve. In this chapter, we acknowledge and address the people, processes, and tools that go into this evolution to help set the expectation of change, and encourage the calculated risks to progress our mobile apps.

Introduction

"I code, therefore, I am." is perhaps a mantra for many of us who really enjoy building mobile or other apps. While most of us are coding for the business, the fact is that developing software for the business ends up being more than just coding. Beyond programming, which is such a personal task, we must resolve defects and test or communicate to testers about fixes; we must update our extensive task lists to project/program management; we must purposefully design and ensure others know and are following the intent; and of course, we must deliver our code into production and be open to feedback from users.

Admit it! At the end of your day or sprint you would have created or touched several other assets beyond your code. One asset example is a specification, WSDL, XML, UML, which you will likely use as IBM studies indicate that 90% of mobile apps use some kind of backend services, data, or "system of record," which are maintained by other developers who publish those same specifications. This collection of assets, as shown in Table 2.1, follow a natural teaming lifecycle from idea to production, and this lifecycle is repeated until the app does not offer enough value to the user anymore.

Therefore, this chapter is not limited to solely the mobile app code, but it evolves the singular mobile developer into an Enterprise Mobile Development Lifecycle. As a mobile developer, you're likely already familiar with agility, as the market suggests that nearly 100% of all mobile efforts claim to use Agile methods. In this chapter, we have applied Agile application thinking to you and your team's evolution into an Enterprise Mobile Development Lifecycle, which in the end, we expect would show how you can grow your and your team's capabilities to deliver better quality mobile code faster.

9

Table 2.1   Developer Assets Quick Reference

Developer Asset, Purpose	Where Asset/Artifact Stored		Shared with . . .
	Open Source/Other	IBM Options	
Source Code or **any file versions** (created by some IDE)	Github, Subversion, CVS, TFS	Rational Team Concert™	Other developers
Defects, Work Items	Spreadsheets Jira	Rational Team Concert Bluemix, Jazzhub (SaaS)	Testers, other developers, project managers
Project Plans, tasks, schedule, reporting	Microsoft® Project Spreadsheets	Rational Team Concert Bluemix, Jazzhub (SaaS)	Project managers, LOB owners
Interactions	Sticky notes Whiteboard	Rational Design Manager/ Rational Software Architect Rational DNG (formerly RRC)	Users, Requirements Analyst, UX Designers
Visual Designs (storyboarding)		Rational DNG (formerly RRC) for comment	Users, Requirements Analyst, UX Designers
Graphics/Stylings files, created by Adobe Suite or others	Same as "Source Code"	Same as "Source Code"	Graphic artists, UX Designers
Unit tests	Same as "Source Code"	Same as "Source Code"	Other developers
Test Plans	Spreadsheets	Rational Quality Manager	Testers, QA Managers, other developers
Test Cases	Documents/Spreadsheets	Rational Quality Manager	Testers, other developers
Test Execution	Spreadsheets and/or homegrown scripts	Rational Quality Manager Rational Test Workbench (automated test scripts)	Testers, QA Managers, other developers
Test Results	Spreadsheets	Rational Quality Manager Mobile Quality Assurance	Testers, QA Managers, other developers, Project Managers
Builds scripts	Same as "Source Code"	Same as "Source Code"	Other developers, Build engineers
Deployment Scripts	Operating system files	UrbanCode® Deploy	Other developers, DevOps/Deployment engineers

Developer Asset, Purpose	Where Asset/Artifact Stored		Shared with . . .
	Open Source/Other	IBM Options	
Backend systems specifications (WSDL, XML, UML, etc.)	Same as "Source Code"	(*) Rational Design Manager/ Rational Software Architect	Other developers, Enterprise Architects, Software Architects
Binaries (EAR, JAR, BAR, WAR, EXE, DLL, APK, IPA)		Rational Asset Manager	Other developers, Asset managers, Deployment engineers
Provisioning Profiles (iOS app-specific "registration information"), Developer Certificates, Push notification IDs	Text file/email	Same as "Source Code" NOTE: While developer certificates are private to a developer, organizations may wish to catalog these instead of just private storage on an isolated developer machine.	

DevOps and Enterprise Mobile Development Lifecycle Overview

A DevOps Approach Is Core to Delivering Client Value

For organizations that are driving a MobileFirst strategy,[1] it ultimately comes down to the overall mobile application experience that will determine success or failure of the Business-to-Consumer (B2C) or Business-to-Employee (B2E) effort. IBM studies reveal the following key metrics:

- 80% of mobile applications are used once and *then deleted.*
- 90% of smartphone users keep their phones with them *continuously.*
- 57% of clients face challenges in delivering a quality experience and the functional capability in mobile applications.
- 50% of outsourced projects are expected to under perform.
- 85% of adults expect the mobile experience to be better than using a laptop or desktop.
- 65% of consumers would not purchase products from your company if they have an initial bad app experience (Forrester Research).

Providing an overall high-quality mobile application experience is highly dependent on a continuous software delivery approach. It is not sufficient anymore to outsource or build the mobile app once and then let it stagnate—mobile apps require not only an initial operational capability but also an ongoing, agile evolution as:

---

[1]http://www.ibm.com/mobilefirst

- User needs evolve

- Business needs evolve

- Competitive apps are releases

- User interface (UI) innovations occur

- The underlying platform evolves

- Other information sources become available for integration including BigData, Social Collaboration, Systems of Record, and managed application programming interfaces (APIs).

DevOps is an enterprise capability for continuous software delivery of mobile applications. To demonstrate market opportunity for DevOps, the Institute of Business Value study reveals that 86% of companies believe software delivery is important or critical, yet, only 25% leverage software deliver effectively today, and nearly 70% of those that do, outperform their competitors who do not leverage this capability. For more detail and best practices on DevOps, see Chapter 7.

Let me offer another perspective: projects can still fail even if project managers expertly manage schedules, analysts diligently capture business requirements, developers write quality code, and testers run thousands of tests. If all of this work is performed in silos, those efforts lead to failure, misunderstandings, and a lack of trust among team members. Even if you have teams that are the best in the world in their discipline, you can still fail if they are not working together and sharing information.

One Essential View of DevOps: The Mobile Developer Perspective

While delivering an overall mobile application experience goes beyond just the code construction of a mobile developer or team of mobile developers, this chapter focuses on the key aspects of delivering great code shown in Figure 2.1.

The process in this diagram is a repeatable lifecycle that allows clients to continuously deliver high-quality mobile apps and rapidly respond to feedback at the rate and pace of the agile mobile market demands. The key steps or capabilities in the lifecycle include:

- Design and Develop: Build enterprise ready applications in a cross platform way.

- Integrate: Easily link to existing data assets or cloud services.

- Instrument: Provides a way client can understand what is happening with the application from a security, usability, and quality perspective.

- Test: Verify the app to produce a great experience and deliver the required quality.

- Scan and Certify: Scan and certify against app vulnerabilities and resolve potential compliance issues.

- Deploy: Ensure that the mobile app is deployed with the right performance and scale.

- Manage: Ensure that app or device has the appropriate management and governance.

- Obtain Insight: The client can examine the user experience and be able to determine how to effect change, and then link that back into a continuous delivery cycle.

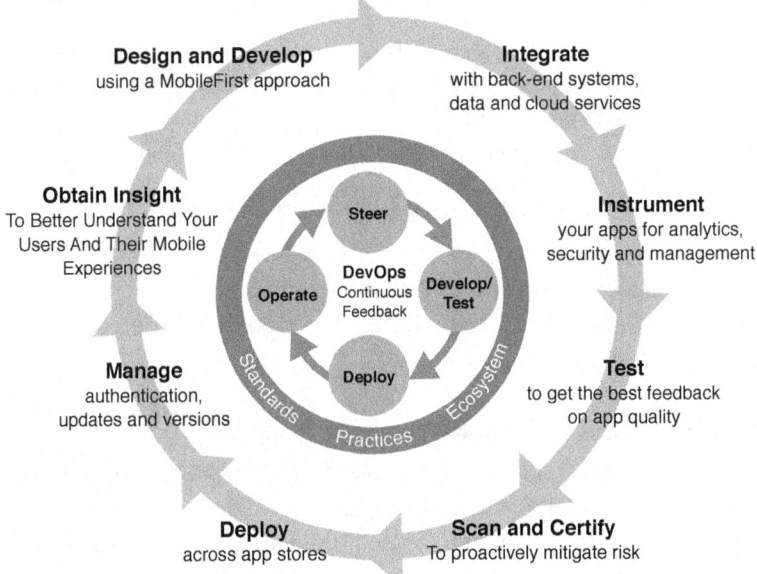

Figure 2.1   Enterprise mobile development lifecycle

While these steps are listed in sequence, the reality of adding more capabilities to your team is that it is very similar to adding capabilities to your application: you iterate your way there and add more capabilities as you progress, evolving the team's capabilities over the life span of the effort. History suggests that attempting to add all of these capabilities at once would be too disruptive for any team as new technologies often involve a learning curve.

The remainder of this chapter takes you through a series of suggested "sprints" of the above capabilities. Please note that this is just a suggestion based on the progression of a mobile app and your actual implementation of capabilities may be different based on your business needs and financial justification. For assistance in prioritizing capabilities and aligning them to your business value, consider a DevOps Assessment.[2]

Sprint 1: "Hello World"—Initial Operating Capability and Prototype

As organizations seek to get to market first, investments are made in small teams of "heroes" that abandon most well-defined institutional processes to build a mobile application. In some of these cases, these investments are "side-projects" or experimental prototypes to help internally "sell" the concept to established business units. Interestingly enough, when these prototypes are created with an optimized user experience, the reaction is almost always like one client's story, "we need that app to help our mobile workforce be more productive." The result of this "selling" now means the effort must be turned into a real project!

---

[2] https://www.ibm.com/developerworks/community/blogs/invisiblethread/entry/diy_with_ibm_s_self_assessment_tool_for_devops_practices?lang=en

Design—The Big Picture

IBM Founder Thomas J Watson quoted, "Good Design is Good Business." Developers and teams like the one mentioned above have to start somewhere, so we provide some tactical advice here, while Chapter 3 addresses this in more detail.

In this first "sprint," we will cover some brief, but different design aspects:

- Coding the basic User Experience (UX) Design—it is where everyone starts!
- Designing and Rapid Prototyping—a continuous activity to meet evolving user need
- Leveraging Solution Design—creating or leveraging the Leveraging Service Oriented Architecture (SOA)/Web Services operational components of the environment in which the application runs

Design: Turning Your Primary User Story into a Basic UI

Whether you are building an API, a mobile app or other platform effort, your first place to deliver valuable software is to really understand what single thing you will deliver as value first to the user: this is your initial operating capability. While prioritizing the many requests and determining value itself can be a detailed line-of-business best practice, those are capabilities we can add later, leaving us to focus initially on building an interface to represent this initial value.

Most projects simply begin with the native toolkits and Software Development Kits (SDKs). Historically SDKs have instructed you to build the UI with code in a "Hello World" type of tutorial. Most often in these beginner tutorials, the goal is to help you get all of the way through to a deployment of your app, so that you can get a feel for the overall process, much like this chapter itself does!

As for the UI, with the Android SDK you build an XML code representation of your UI, which is described in the "Building a Simple User Interface" of the developer training.[3] Tools in this basic UI category, however, are advancing quickly, and depending on your choice of platforms, the experience can be vastly different. For iOS developers, you incorporate your code into storyboards and scenes that you drag and drop, which is a good lead-in to the next section.

Design: UI Mock-ups and Rapid Prototyping

IBM's Design Group[4] suggests these best practices of Design Thinking. In the end, it really comes down to:

- "Look": Visual appearance including but not limited to the UI components
- "Behavior": Focused on the process or flow of the user interaction

While there are a number of free (e.g., http://maqetta.org) and low-cost products (e.g., http://uxpin.com/, http://balsamiq.com) in the market for building UI "mock-ups" or "wireframes," IBM's mobile, Eclipse-based IDE's have a common UI builder: the Rich Page Editor. Based on

---

[3] http://developer.android.com/training/basics/firstapp/building-ui.html

[4] http://www.ibm.com/design/

a WYSIWG approach, the **Rich Page Editor** not only creates the UI using an easy drag and drop approach, but also provides guidance on the placement of typical UI elements (i.e., buttons), and generates the code dynamically for the interface. In fact, the Rich Page Editor supports a WYSIWG design view, source view or a split view that combines the Source and Design views in a split screen view so you can see the changes from one view reflected in the other. The Android SDK has a similar "Design/Preview" concept, for your manual XML coding (Figure 2.2).

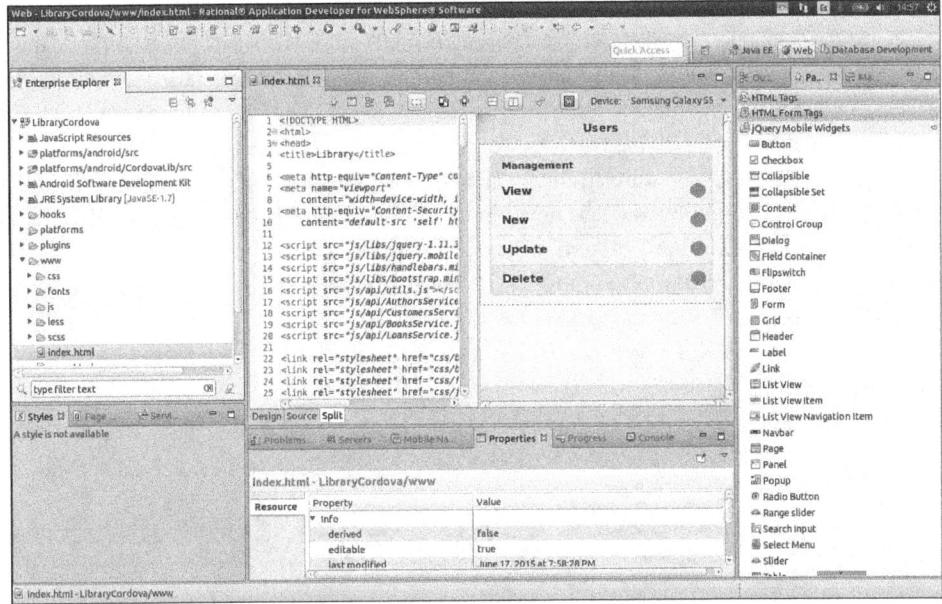

Figure 2.2   Rich Page Editor supports WYSIWIG design

The Rich Page Editor can be found both in the MobileFirst Platform Developer Edition[5] and the Rational Application Developer (RAD) IDE.[6] The choice of IDE is not mutually exclusive for enterprise applications—refer to the following articles for more information and guidance on IDE selection:

- Server-side mobile application development: Part 5. Integrate the IBM MobileFirst/ Worklight® adapter with EJB application[7]

- Take full advantage of IBM's IDEs for end-to-end mobile development[8]

---

[5] http://www.ibm.com/software/products/en/mobilefirstfoundation

[6] http://www-03.ibm.com/software/products/en/application

[7] https://www.ibm.com/developerworks/community/blogs/e4210f90-a515-41c9-a487-8fc7d79f61/entry/ part_5_integrate_the_ibm_worklight_adapter_with_ejb_3_1_application?lang=en

[8] https://www.ibm.com/developerworks/community/wikis/home?lang=en#!/wiki/2fad2df2-9c68-4aa3- abba-01e910211998/page/5531e600-93c7-46f6-86f4-a570b1bcf391/attachments

One distinct IBM advantage for this Rich Page Editor capability is this design/source view showing code as you create the interface. Many alternatives on that market that produce mock-ups, later have to be given to developers as a work product to help developers separately code the solution.

There have been *integrated* partner solutions for prototyping, such as iRise Connect for IBM® Rational[9] that makes high-fidelity visualizations instantly accessible from within IBM tools. Through the history of application development, many advances have been made to integrate the benefits of rapid prototyping into actual reusable application components and once can expect the market to see more convergence in this area as more innovations meet standards for development.

As for the "behavior," while often "reverse engineered" via UI construction, a common mistake is to build an interface without clearly understanding the user interaction first. Whether you use a whiteboard or a tool, we recommend a standards-based approach like UML use case or sequence diagrams, as the standard is well-defined on how we can capture and document such interactions.

Tools like Rational Software Architect (RSA)/Design Manager,[10] implements UML 2 and the typical diagrams found in use case development: Use Case, Scenario, Activity diagrams, as well as "agile sketch diagram" support for more informal information capture (which can be formalized later if required). If you opt for textual user stories, you can capture these in Rational Team Concert[11] as part of your sprint planning (Figure 2.3).

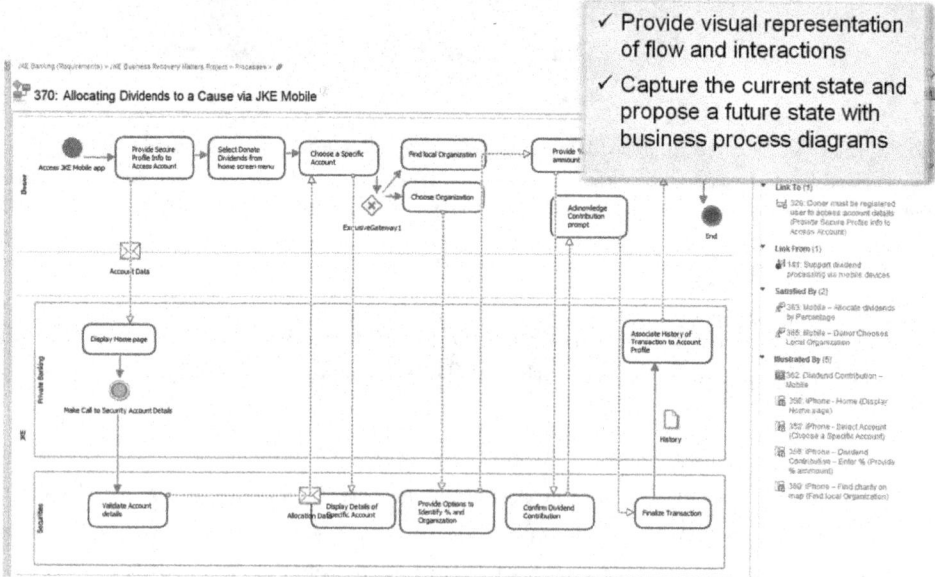

Figure 2.3   Diagrams communicate interactions and flows for posterity

---

[9] http://ww2.irise.com/resources/view_video/generalVideo_28

[10] http://www-03.ibm.com/software/products/en/rational-software-architect-family

[11] https://jazz.net/products/rational-team-concert/

We mention both of these early because user experience is critical and documenting this for your future team's understanding can really improve productivity just like the IBM Interactive team benefits[12] when they bring on new team members and use UML to streamline communication. Similarly, if your enterprise data and services teams have completed their SOA or other designs using RSA Design Manager, then you would be able to integrate this Solution Design information more quickly into your mobile application, which is a good place to mention this as we transition to this very topic.

Integrate: Enterprise Services and Data

As mentioned previously, 90% of mobile applications leverage backend enterprise data and services, and therefore integration to these systems and the developers that support those systems is important to continuous software delivery. Despite this fact, in working with several clients and IBM Business Partners, we are finding that there is still a significant portion of companies that have not invested in publishing those services nor have they made a transition to SOA to make the transition to mobile easier.

If you have not yet made that transition, please refer to Chapter 5 for approaches leveraging the Enterprise Service Bus, DataPower®, CastIron®, and other mechanisms for integrating enterprise services and data.

If you have already made that transition, you may be in the position to leverage your existing Enterprise Services and Data designs, or in some cases even generate code for your mobile client based on existing RESTful services that you might have catalogued already in your enterprise design (Figure 2.4).

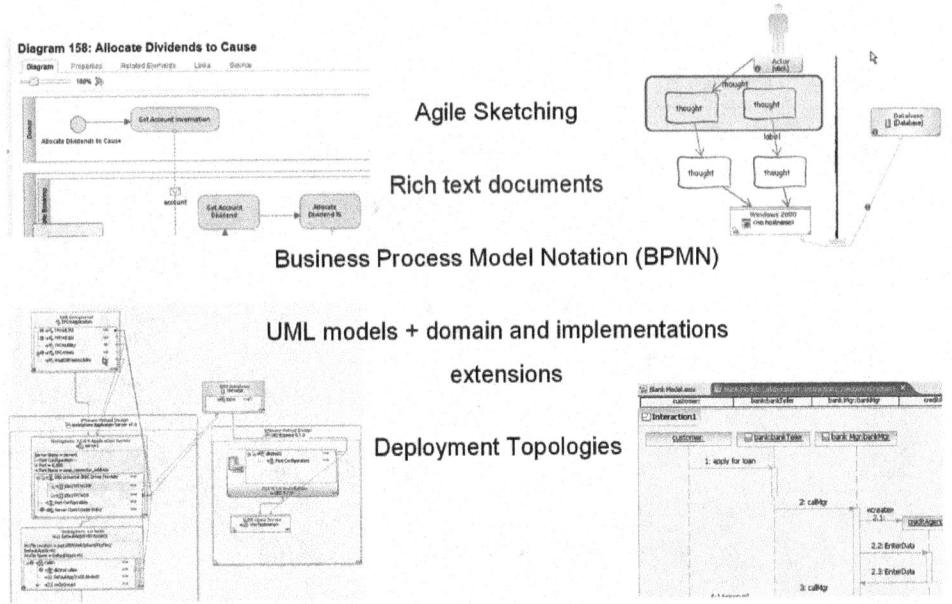

Figure 2.4 Existing standard designs accelerate enterprise integration

---

[12] http://www-03.ibm.com/software/businesscasestudies/no/no/corp?synkey=A951466Y90458U66

Test: Manual Testing Is Better than No Testing

Nearly all mobile teams use manual testing as their primary approach for verifying quality of the application. This isn't really a surprise, as history tells us that test automation, as described in Chapter 6, doesn't often keep up with the developer innovation, but as the market matures that will be a different story. For initial operating capability, testing is specifically aimed at verifying quality of that single piece of functionality that is providing value to the client. At this point in the cycle, there is not much else to test, manage, or provide any significant regression tests upon, so only manual testing makes sense here. The primary advice here is to not only do some manual testing, as some teams even overlook this, but also to start to get the testing organized with clarity on what to test, whether it is sharing out the single design or prototype with the people testing, or constructing some simple test procedures for the same audience.

Test: Simulate/Preview

In these early stages, developers are typically the ones also doing the testing efforts. One developer-focused manual approach to visual verification of the application is using IBM MobileFirst Platform Developer Mobile Browser Simulator. While technically not a test, the Mobile Browser simulator does give the developer a rapid application development/prototyping approach to building the UI and then reviewing the apps look and feel in a variety of different platform configurations, including a change in orientation (Figure 2.5).

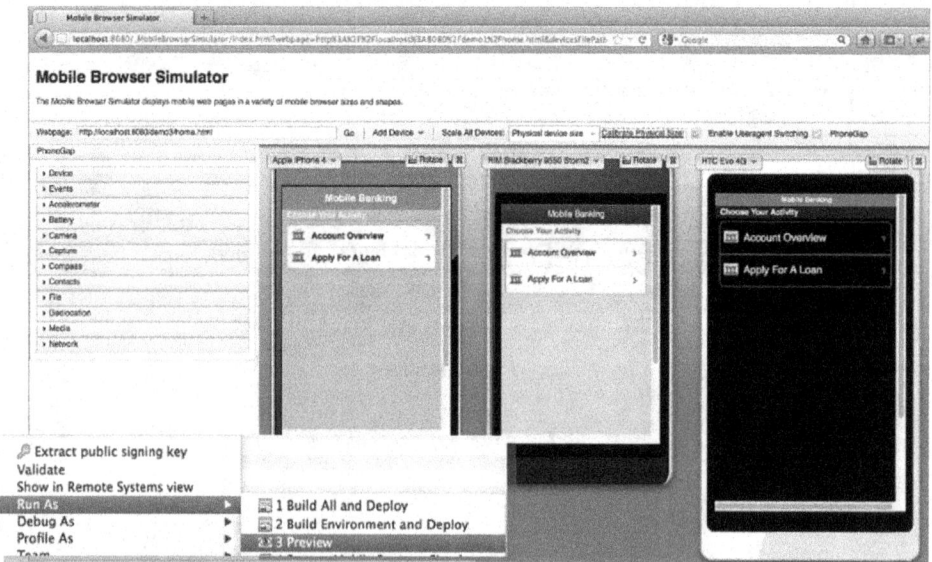

Figure 2.5 Mobile preview capabilities

Sprint 2: "Hello DevOps!"—Improve Developer Productivity

Welcome to the first chapter that really goes beyond the coding experience and looks to improve developer productivity with additional approaches and tools to help developers continuously delivery higher quality mobile applications.

Develop: Productivity with Wizards and Mobile Platform Portability

The MobileFirst Platform Developer Edition offers a more productive set of design and development features. Use wizards, visual UI editors (like the Rich Page Editor discussed previously), prepackaged UI frameworks, and JavaScript API for common UI controls. Apache Cordova is prepackaged with MobileFirst Platform.

Project creation wizard creates the initial structure for the application, including packaging structure for UI and logic components, backend adapters, and application deployment to MobileFirst Platform server. To quickly help target new mobile platforms, you can add platforms "on-the-fly" (while you continue to promote code reuse), and the MobileFirst Platform Developer Edition creates the necessary project structures for Android or iOS' XCode automatically, streamlining your ability to build your native or hybrid application.

Instrument: Quality Assurance and Testing "In the Wild"

IBM's Mobile Quality Assurance (MQA)[13] improves application quality and the customer experience through streamlined reporting of defects, automated crash reporting, sentiment analysis, and improves turnaround time on fixes with over-the-air distribution of new builds (Figure 2.6).

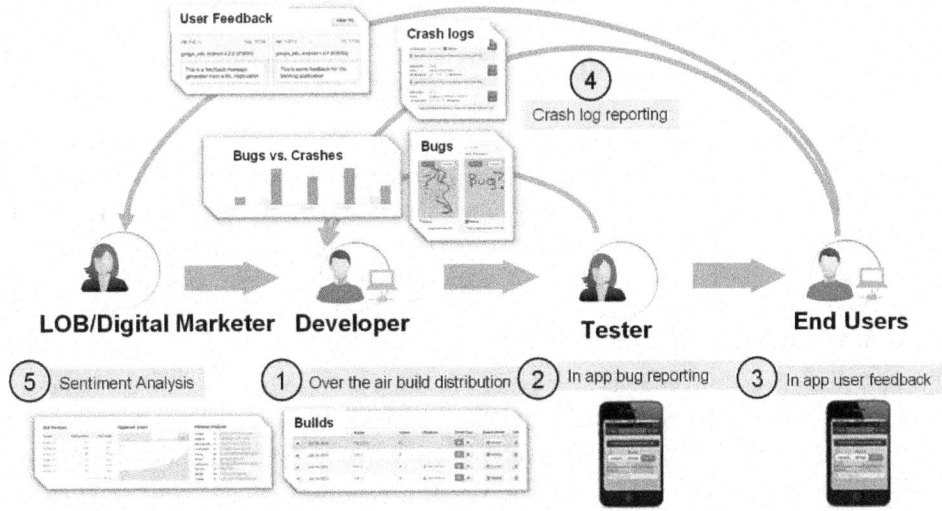

Figure 2.6 Key value scenarios for Mobile Quality Assurance (MQA)

---

[13] http://www.ibm.biz/mobilequalityassurance

Defect reporting is easy with IBM MQA: simply shake the phone to initiate a bug report and a screen shot is captured along with several important pieces of information about the phone state (i.e., OS, version, manufacturer, signal strength). Typically, capturing this information and manually entering it into a tracking system may take about 10 to 15 minutes per defect and is often time-prohibitive for testers to gather all of this information to provide quality analytics to the development team. From a time savings return on investment, just 150 bug reports is one week of mobile team time saved (Figure 2.7). Once captured, crashes or bug reports can be analyzed by these characteristics, such as platform or OS version, to determine if quality issues are specific to a particular mobile environment—something that otherwise, would be very difficult to identify until it is too late and the user uninstalls the app.

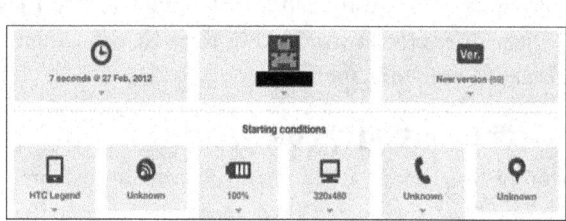
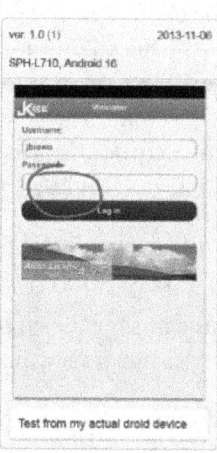

Figure 2.7   Sample bug report from Mobile Quality Assurance

To take advantage of this capability, simply instrument, add the MQA SDK calls into your mobile app per the online help instructions, and you can begin to take advantage of these time saving capabilities. These changes take less than an hour to do and you only need to do them once. Libraries exist for both preproduction testing and an even lighter weight production library can be used to deploy the app into app stores, enabling you to collect user feedback and crash reporting.

Test: Code-Centric and Unit Testing

JUnit and Selenium are two code-centric types of testing mechanisms—you must write additional test code to execute these tests. The Selenium project automates browsers, primarily for automating mobile web applications for testing purposes. Selenium has extended from mobile web apps into native and hybrid app testing with Selendroid and iOS-driver (neither as of the date of this entry have reached version 1), which uses the Android instrumentation framework, and UI Automation framework from Apple, respectively. While there is a Selenium IDE that plugs into the browser to help export recorded tests to code, the Selenium documentation states:

We don't, however, recommend you do all your test automation using Selenium IDE. To effectively use Selenium you will need to build and run your tests using either Selenium 2 or Selenium 1 in conjunction with one of the supported programming languages.[14]

Clients who are already using this approach would find additional productivity value in these IBM tools. Each of these is described more fully in the coming sections because these may involve professional testers that do not necessarily have programming skills applicable to developer-centric testing:

- Rational Test Workbench—provides record and playback of automated tests with no coding required.

- Rational Quality Manager (RQM)—plans, designs, reviews, and executes a variety of different tests including but not limited to JUnit Selenium tests and tests created in Rational Test Workbench.

- To support DevOps Continuous Testing, most any test can be included as part of an automation workflow and further enable continuous delivery activities using IBM UrbanCode Deploy.

Deploy: Automate Your Deployment Pipeline

While "Release and Deploy" most often addresses the code release and deployment of a software solution to test and production environments, at its core, this approach is an automation step that also applies to various middleware components, such as Messaging, Enterprise Service Bus, as well as the Application Servers, where often times manual steps can be improved with automation for more reliable and consistent results.[15]

Continuous release and deployment provides a continuous delivery pipeline that automates deployments to test and production environments. It reduces the amount of manual labor, resource wait-time, and rework by means of push-button deployments that allow higher frequency of releases, reduced errors, and end-to-end transparency for compliance. IBM's products for Continuous Release and Deployment:

- IBM® UrbanCode Deploy[16] orchestrates and automates the deployment of applications, middleware configurations, and database changes into development, test, and production environments (Figure 2.8). While UrbanCode Deploy integrates with many open source tools that are found in clients today, it also adds additional client value with integration to IBM MobileFirst Platform Developer to streamline mobile apps into deployment. Please refer to the UrbanCode Plugins page for integrations with middleware. To support DevOps Continuous Testing (testing describe in the next section), almost any

---

[14] http://docs.seleniumhq.org/docs/01_introducing_selenium.jsp#choosing-your-selenium-tool

[15] https://developer.ibm.com/urbancode/plugins/

[16] http://www.ibm.com/software/products/en/ucdep/

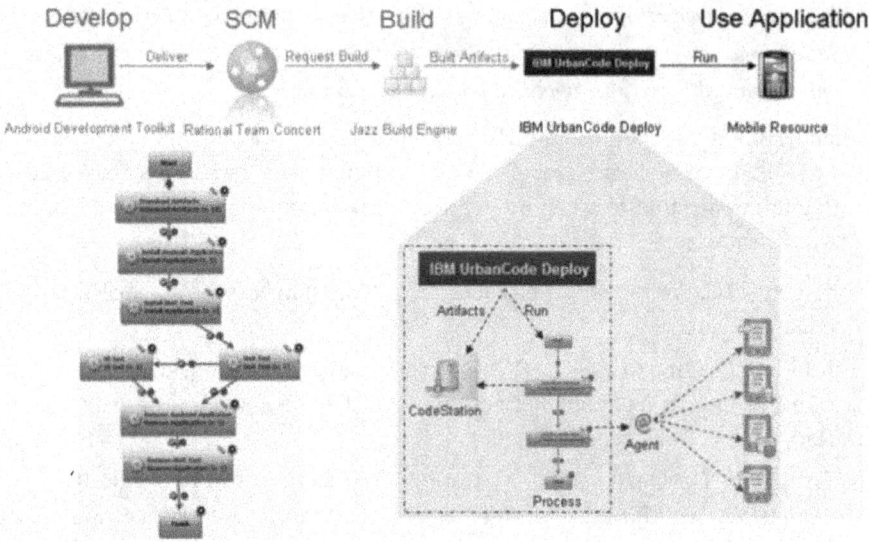

Figure 2.8   Improving the Android mobile delivery pipeline with IBM UrbanCode Deploy

test can be included as part of an automation workflow and further enable continuous delivery activities using IBM UrbanCode Deploy.

- IBM MobileFirst Platform Application Center[17] securely deploys apps through a private enterprise app store (Figure 2.9). Developers can distribute mobile builds and elicit feedback from development and test team members.

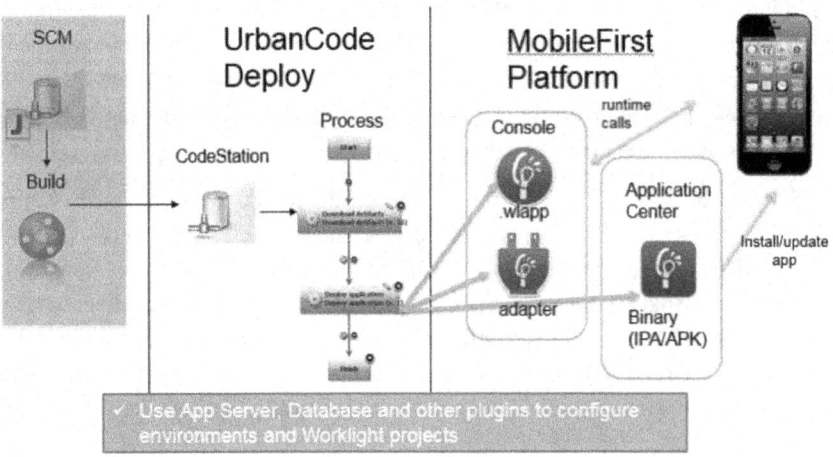

Figure 2.9   IBM UrbanCode Deploy integration with IBM MobileFirst Platform

---

[17] https://www-01.ibm.com/support/knowledgecenter/SSHS8R_7.1.0/com.ibm.worklight.getstart.doc/start/c_wl_overview.html

- IBM MQA[18] provides an "over-the-air" distribution model for application updates under test, where specific versions can be selected to be deployed to select users. API-level distribution control is also provided to support continuous integration or continuous deployment activities as shown above.

- IBM® UrbanCode Release[19] manages the release of complex interdependent applications, infrastructure changes, and simultaneous deployments of multiple applications.

- IBM Cloud Orchestrator[20] provides an open and extensible cloud management platform for managing heterogeneous hybrid environments. The software integrates provisioning, metering, usage, and accounting as well as monitoring and capacity management of cloud services.

Sprint 3: "Software Delivery Is a Team Sport!"

Now that we have improved individual developer productivity, let us focus on expanding our success beyond ourselves, after all, a (small) team will produce more than an individual. This is where things begin to get interesting from a changed perspective, as high-performance teaming usually follows a "forming," "storming," "norming," "performing" model.[21]

Develop: Agile Planning

Nearly all mobile development teams use Agile processes and planning. IBM DevOps Services for Bluemix and Rational Team Concert supports sprint planning and backlog management, and captures information about the work items as developers complete them (Figure 2.10).

Develop: Work Items (Defects, Enhancement Requests)

All developers fix defects and implement enhancement requests, but how developers take on these work items varies greatly from organization to organization. As software teams or departments grow, they will quickly outgrow the email and spreadsheet tactical implementations. Mostly gone are the days of home-grown defect tracking systems, as open source solutions such as Jira become more visible. But this is a market with 100's of no/low-cost tools for the specific task of tracking some data about a task through some workflow, and implementations of data field security (who can change what data fields), and integrations also vary tremendously among the various market offerings.

IBM MQA, mentioned previously, already has integration into Rational Team Concert or IBM DevOps Services for Bluemix to streamline defect or feedback entry into your backlog.

---

[18] http://www.ibm.biz/mobilequalityassurance

[19] http://www.ibm.com/software/products/en/ucrel/

[20] www.ibm.com/software/products/en/ibm-cloud-orchestrator

[21] Tuckman, Bruce W. 1965. Developmental sequence in small groups, *Psychological Bulletin*, 63, 384–399. https://en.wikipedia.org/wiki/Bruce_Tuckman

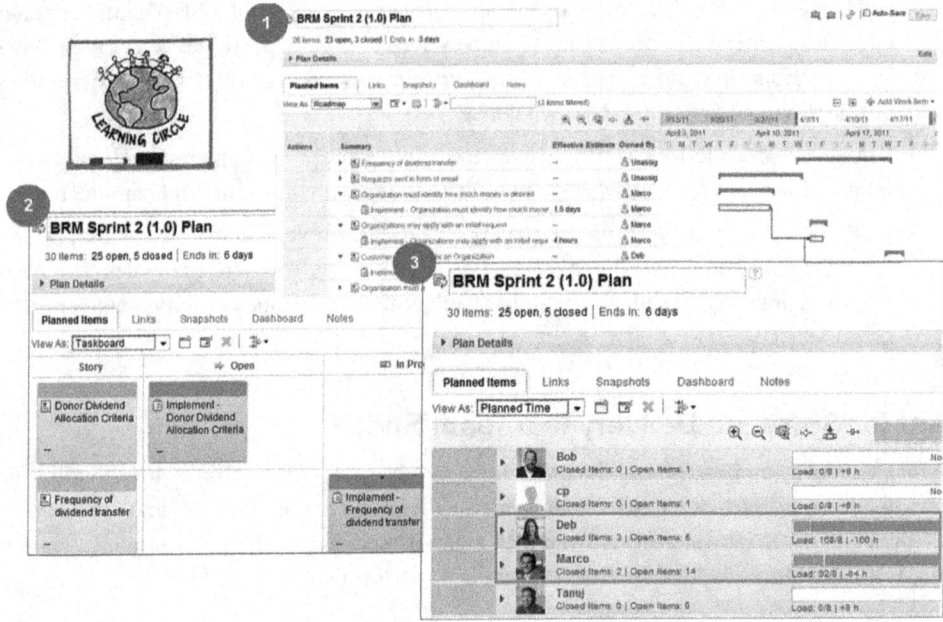

Figure 2.10   Accelerate time to delivery with Real-Time Planning

Develop: Application Lifecycle Management (ALM) Imperatives

As you progress from developer to team productivity, you absolutely need a common vision and mission to unite the team. Teams should, therefore, discuss a set of imperatives that increase productivity through a common approach across all roles of the project.

From this reference,[22]

> *"many organizations are faced with hastened delivery schedules due to competitive pressures and the need to innovate. Yet software development is difficult, and the software systems that are maintained and delivered by the world's IT and device development organizations are astoundingly complex. Teams challenged by reduced time to delivery must do so without increasing their budgets or sacrificing quality. Their strategy, instead, must be to improve software development efficiency. A solution to this dilemma is to improve Lifecycle Collaboration with Application Lifecycle Management.*
>
> *Designed for the execution of a software delivery project, Application Lifecycle Management solutions coordinate people, processes, and tools in an iterative cycle of integrated software development activities, including planning and change management, requirements definition*

---

[22] https://jazz.net/library/article/637

and management, architecture management, software configuration management, build and deployment automation, and quality management. In addition to the capabilities, the fundamental features of an ALM solution include traceability across lifecycle artifacts, process definition and enactment, and reporting."

This same article also states that there is no "one-size fits all" solution, and for mobile there appears to be more focus on Agile/Real-time planning and these two imperatives briefly described here (for the two other imperatives, see https://jazz.net/library/article/637):

- Establish Lifecycle Traceability of Related Artifacts—developers may often state, "it would be nice if I could see all the tests associated with this defect" or "what builds are associated with a work item?" For mobile development teams, this information should be at their fingertips so that development and design analytics[23] can help individuals find the information they need quickly (Figure 2.11).

- Enable In-Context Collaboration—having a single source of truth where project information resides, this imperative avoids relying on emails, chat programs, disconnected spreadsheets, and "word of mouth" as collaboration tools.

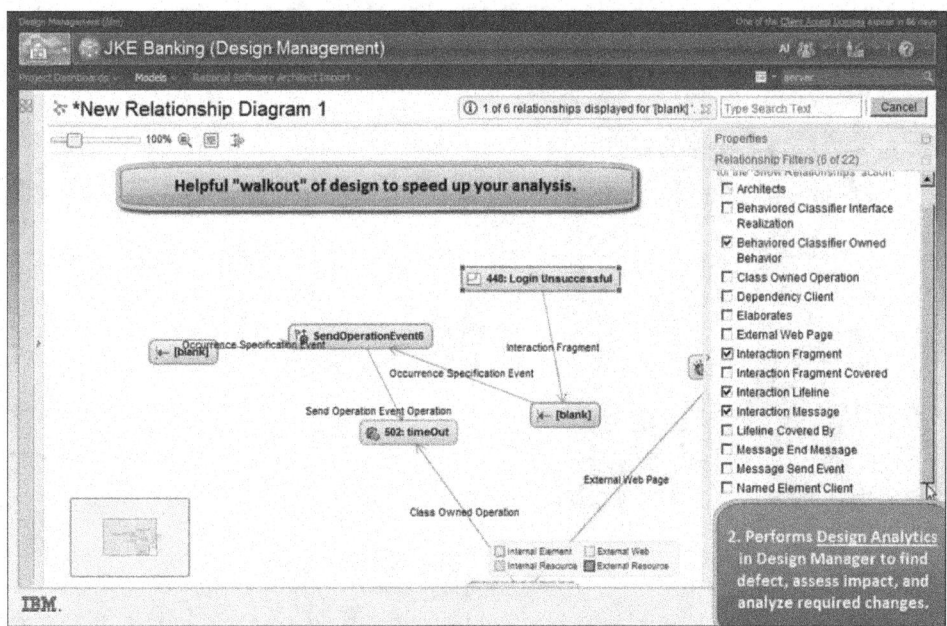

Figure 2.11 Lifecycle Traceability of related artifacts helps team members easily find information

---

[23] http://www.youtube.com/watch?v=NovQGpa2H6E

Test: Improving Manual, Automating User Interface Tests, and Test Data Management

This section is purposefully short as there is a detailed testing Chapter 6 later in the book. Teams should not overlook improving the busy "office work" that often occurs with planning and organizing manual testing efforts. While many projects strive for "100% automation," teams rarely ever get to 85% automation. Therefore, manual testing is here to stay. RQM can execute a variety of different tests including but not limited to JUnit Selenium tests and tests created in Rational Test Workbench. Studies show that most clients are still doing a majority of their testing manually, and RQM can help automate many of the tasks of planning, designing, and executing manual tests. Furthermore, there is a mobile app for RQM[24] that helps clients run manual tests (this approach is not limited to testing mobile apps).

Automated UI testing is often the next place mobile teams pursue, primarily because of the vast productivity gains typically associated with a "record/playback" type of approach, which can run tests more consistently and during any hour of the day to "verify quality" of your application's UI. Rational Test Workbench provides this record and playback of automated tests with no coding required (Figure 2.12). Recording can occur on the device itself or in an emulator, and produces a human readable set of commands that you can customize through the RTW interface. This reduces the need to have developers do all of the testing and development, and can shift testing to QA professionals who may not have the development experience to use code-centric testing approaches.

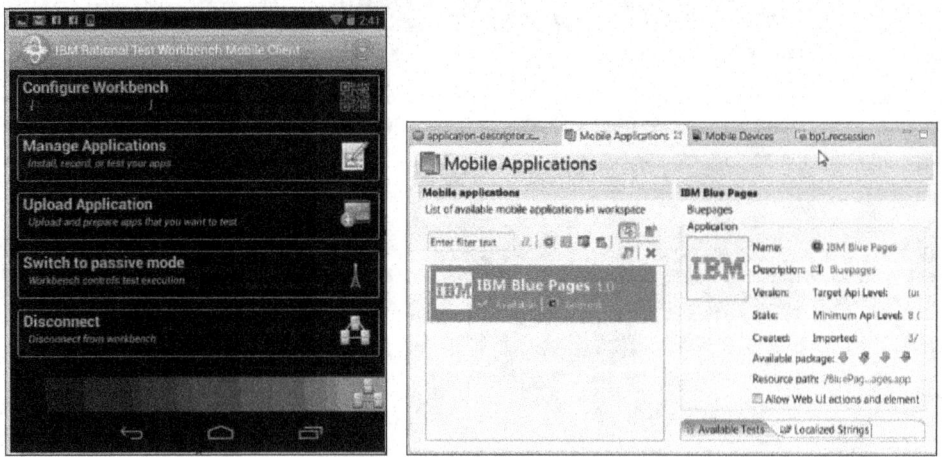

Figure 2.12   Rational Test Workbench Mobile Client (left) and Test Console (right)

As you will see later, an Eclipse environment instruments an application to the targeted device running a thin mobile client and drives the UI automation. These kinds of tests can run locally or in a device cloud, such as IBM Business Partner Keynote's DeviceAnywhere.

---

[24] https://www.youtube.com/watch?v=Tm9MEk8dZzA

Mobile teams may not have time to do manual testing with multiple data conditions and once again, automation is essential in improving the mobile application experience. IBM® Info-Sphere® Optim™ Test Data Management[25] optimizes and automates the test data management process. Prebuilt workflows and services on demand facilitate continuous testing and Agile software development. IBM InfoSphere Optim Test Data Management helps development and testing teams use realistic, right-sized test databases, or data warehouses to accelerate application development. InfoSphere Optim Test Data Management helps organizations:

- Streamline test data management processes to help reduce costs and speed application delivery.

- Analyze and refresh test data on demand for developers and testers.

- Create production-like environments to shorten iterative testing cycles, support continuous testing, and accelerate time to market.

- Protect sensitive data based on business polices and help reduce risk in testing, training, and development environments.

- Use a single, scalable enterprise solution across applications, databases, and operating systems.

- Provides a comprehensive continuous testing[26] solution through Rational Test Workbench[27] for functional, regression, integration (service virtualization), and load testing.

Sprint 4: "Ruggedized for the App Store"

Now that you have gotten your team involved, together, you will want to "ruggedize" your mobile app for the app store, that is, you'll want to verify quality and security concerns before you release something that may cause a negative brand impact and/or poor ratings in the public app stores. For private or enterprise-level app stores, the same applies and in fact, your organization may already have some service-level agreements in place for performance or security standards that you should meet prior to publishing to the internal app store.

Test: "FURPS" and Virtualization

When testing, there are several dimensions you should consider in this process, which are well categorized by the acronym "FURPS." Each of these dimensions is related to the kinds of requirements that often drive acceptability of any application (see the subsequent Chapter 6 on testing for more detail):

- Functionality—verifying system functions through UI testing or protocol analysis
- Usability—verifying the user experience either manually or through UI testing

---

[25] http://www.ibm.com/software/products/en/infosphere-optim-test-data-management/

[26] http://www.ibm.com/ibm/devops/us/en/

[27] http://www.ibm.com/software/products/us/en/rtw

- Reliability—verifying the stability and uptime, through continuous automation testing or stress testing

- Performance—verifying the system meets performance service level agreements through scalability testing

- Supportability—representing the serviceability or maintainability of the system under test

Expanding this definition, see the Wikipedia page on FURPS and FURPS+, you might push back and consider that all of this testing will take time, leading you to the conclusion that these additions could not possibly keep you delivering software continuously! To some extent, testing and good practices are your "insurance" against some of the risk factors that arise in most software projects. Agility is important to your mobile project and testing more frequently with automation is an approach to address that risk.

Test virtualization is a technique that can boost a mobile team's ability to test more frequently as it often takes time to set up a backend test environment, not to mention the costs usually associated with a test environment. Virtualization starts by capturing the actual traffic between the mobile app and its associated enterprise or cloud services. Once identified, then those services can be "stubbed out" of the testing and placed in virtual services that the mobile team can stand up and have complete control to execute tests to match the ongoing spring delivery schedule.

Scan and Certify

The IBM MobileFirst Reference Architecture chapter on Security states:

> *"Secure mobile development practices should be integrated in each stage of the software development lifecycle. Mobile application design and development should be performed with a security conscience and testing teams need to automate their security-specific unit tests early and often during the development lifecycle. This process will help avoid discovery of critical security issues later in the development process and avoid delay of the Mobile application to the market."*

To help our clients "build security in" from the early stages of development, AppScan® Source can be used to scan the source code for vulnerabilities (Figure 2.13). AppScan's other components also help assess existing web sites and web services for vulnerabilities.

Mobile apps may be increasingly leveraging open source software (OSS). While OSS is free, it comes with license obligations that must be met. Some clients may be interested in certifying their source code for compliance through the Ready for IBM Rational business partner solution.[28]

Obtain Insight: Application Quality Feedback and Analytics

In both preproduction and production, developers and line of business/digital marketers can gain insight into the feedback that users and testers provide. This insight can come from a mobile app

---

[28] https://www-304.ibm.com/partnerworld/gsd/showimage.do?id=23066

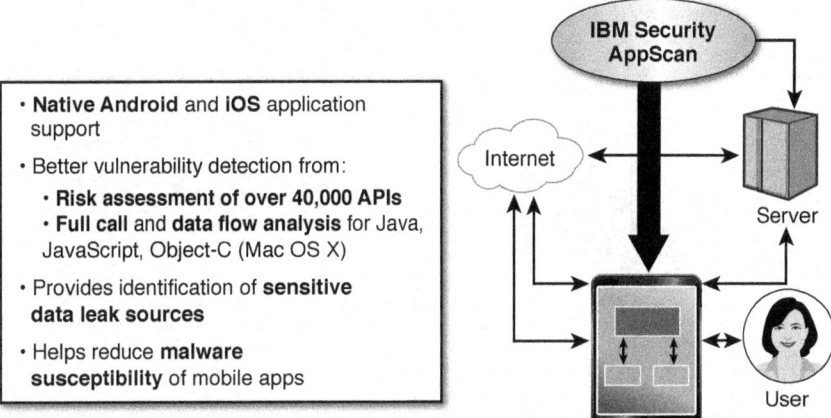

- **Native Android** and **iOS** application support
- Better vulnerability detection from:
  - **Risk assessment of over 40,000 APIs**
  - **Full call** and **data flow analysis** for Java, JavaScript, Object-C (Mac OS X)
- Provides identification of **sensitive data leak sources**
- Helps reduce **malware susceptibility** of mobile apps

Figure 2.13 IBM Security AppScan: identify vulnerabilities in web and mobile application source code

that has been instrumented with IBM's MQA (see the Instrument section earlier in this chapter, Figure 2.14). Preliminary findings suggest that collecting private feedback while successful transactions occur during the mobile application usage may increase the public rating of an application in the app store.

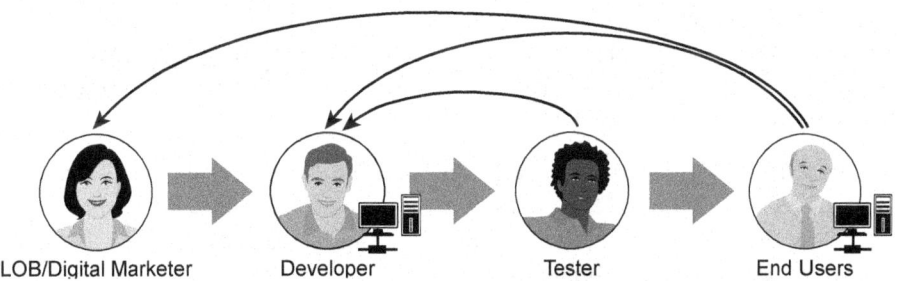

LOB/Digital Marketer   Developer   Tester   End Users

Figure 2.14 Mobile Quality Assurance supports many users

Sprint 5: "Optimizing Enterprise DevOps"

As this chapter comes to a close, our last "sprint" is aimed at optimizing the approach by integrating the enterprise developers and managing the postapp store delivery customer experience and campaigns.

Integrate: Enterprise Developer Integration

Once again, as 90% of mobile applications rely on enterprise data and systems, some level of coordination should be occurring between mobile developers and enterprise developers. In

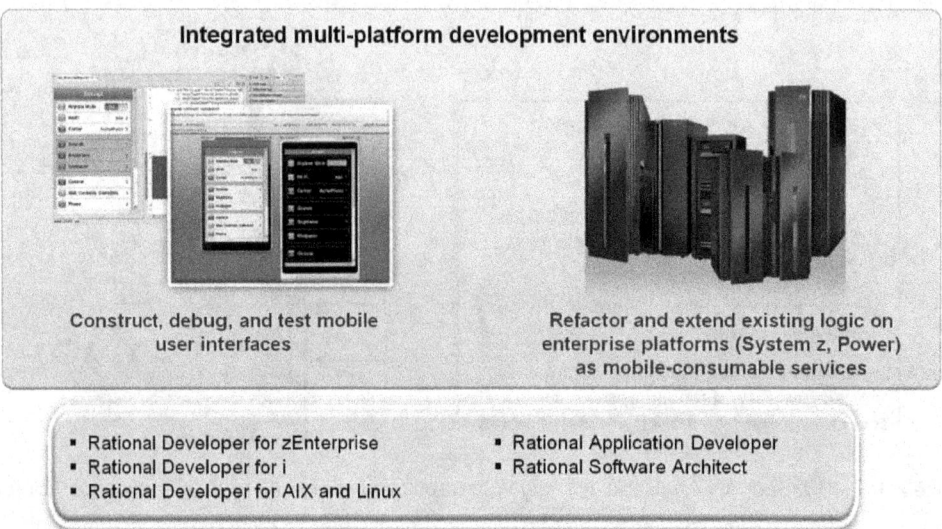

Figure 2.15   IDEs are optimized for the developer on each platform supporting the mobile user experience

addition to mobile UI developers leveraging IBM MobileFirst Platform Developer Edition, the developers working on the enterprise systems can use any of the IBM Rational IDEs to refactor or extend existing systems supporting these platforms: z™ (RDz), Power (RDPower), and WebSphere (RAD or RSA) (Figure 2.15).

Developers need to coordinate work across mobile and enterprise services to ensure overall user experience system behavior. In any particular coordinated release requiring changes on mobile and enterprise systems, you will likely have:

- Work Items—to assign out, track, and manage the tasks across all the platforms involved. One common element to all of the IDEs mentioned above is that they are all Eclipse-based and plugins like Rational Team Concert can help organize the "developer inbox" across those platforms.

- Delivered Code, Integrated Code—as developers contribute, continuous integration, build support, and integration testing will be required to ensure system behavior. Rational Team Concert, which runs on multiple platforms, provides build engines to support this at not only the individual level, but at integration build time as shown in the graphic below (Figure 2.16).

- Tests That Verify the Quality—as described in the Testing chapter (Chapter 6), testing of this integration is vital and Rational Integration Tester as a part of Rational Test Workbench can verify quality of these developer changes.

Rational Team Concert (RTC) Collaborative Development **build engine** provides developers more productivity with:

• Controlled builds

- Individual and

- Team integration and

- Distributed builds for mobile apps

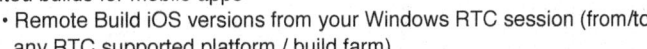

Mobile build server farm

• Remote Build iOS versions from your Windows RTC session (from/to any RTC supported platform / build farm).

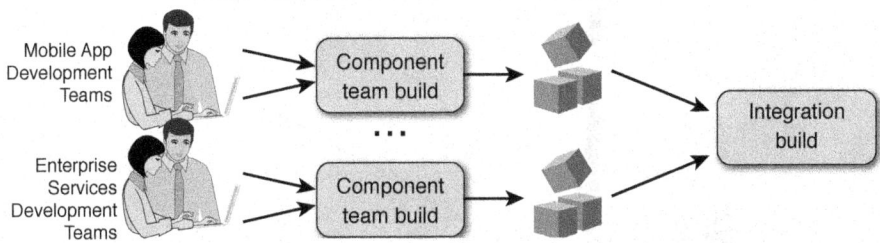

Figure 2.16   Sustain Agility with continuous integration across the enterprise

Instrument and Obtain Insight: Customer Experience (CX) and Campaign Management

Developers can instrument their applications to provide a richer and better quality mobile app experience and get answers to questions like these:

- What parts of the application are users using the most (and perhaps should test the most)?

- What parts of the application are users using the least (or at all—perhaps this code can be eliminated?)?

- Are users not able to complete certain transactions?

Instrumentation involves injecting source code, using APIs from various libraries, into the application, typically manually; however, some instrumentation is automated. These instrumented points send "signal flares" to your server collecting information about your application's usage.

TeaLeaf® uses the client experience information to analyze the customer experience from a business perspective—is the customer able to complete their transaction with minimal frustration and clicks—and application monitoring uses the client experience information to determine if any IT problems in the end-to-end application flow are negatively impacting the customer experience (Figure 2.17).

Xtify® is a mobile campaign management solution for marketers and delivers an SDK that MobileFirst Platform developers can embed in their mobile applications so that mobile marketers can target their app users. Combined with MobileFirst Platform, there are capabilities for push notification, geo-location, and more.

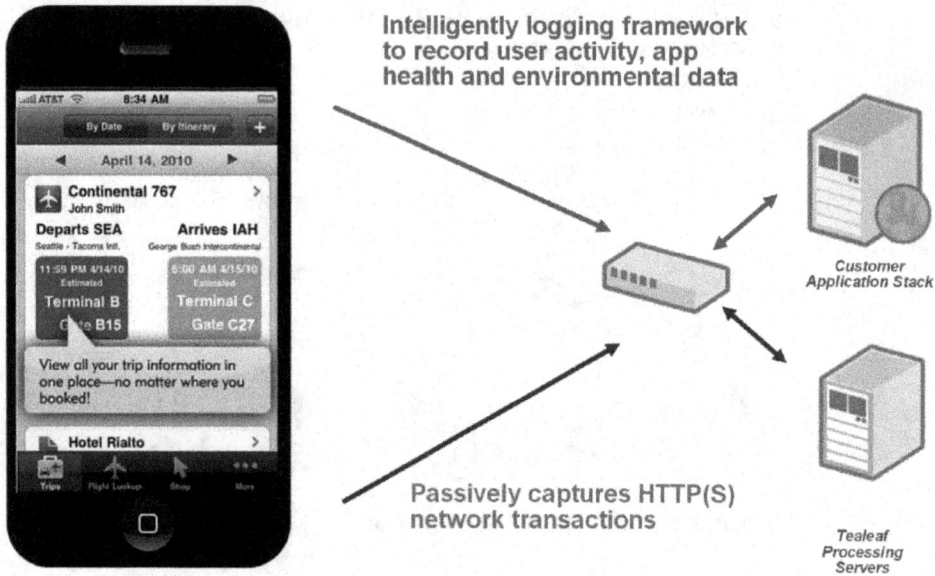

Figure 2.17   Tealeaf native mobile app solution architecture

Obtain (App Store) Insight: Mobile Quality Assurance

While customer experience and marketing campaign management offer private experience data, app stores (Apple App Store and Google Play) offer some quality metrics that can offer insight into public user sentiment. Globally, two new reviews or ratings are entered into app store daily. This information can be harnessed without any instrumentation with MQA to review rating trends or even competitive analysis (Figures 2.18 and 2.19).

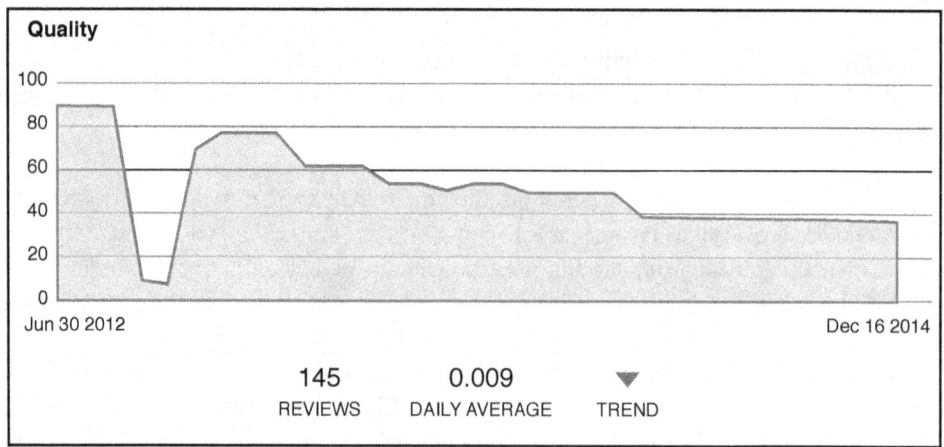

Figure 2.18   Mobile Quality Assurance user sentiment: app rating trends

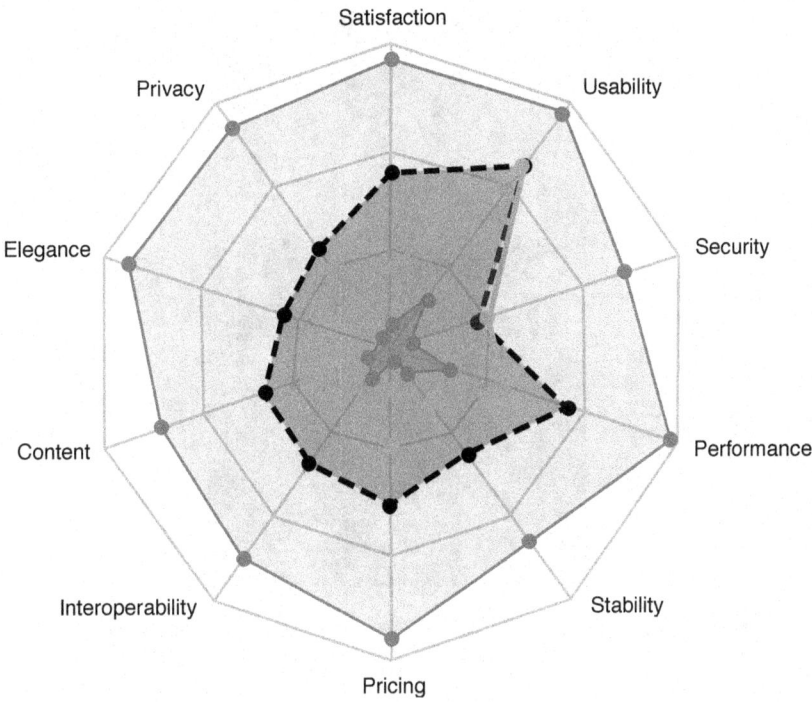

Figure 2.19   Mobile Quality Assurance competitive app quality

Manage: Application Versions, Updates, and More!

Now that you have deployed an app to an app store, and assuming success, you have now deployed your application to thousands (hopefully more) of devices or endpoints. While there are a multitude of considerations for the operational environment discussed in Chapter 7, you as developer have to manage your versions carefully moving forward. You have experienced that there is a built-in delay in getting apps published. There may be times when you require a critical update but cannot afford that built-in delay. If you have built a hybrid application, the MobileFirst Platform provides a "Direct Update" feature that allows you to quickly update application web resources (HTML, JavaScript, and CSS) without going through the vendor (Apple/Google) app store review process (Figure 2.20).

When you deploy the latest build without changing its version to the MobileFirst Server, the next time the app tries to access the server, it will automatically retrieve the latest web resources after prompting the user to accept the update. Direct Update cannot be used to update native code.

For those building internal enterprise applications, The MobileFirst Platform Application Center can be used as an Enterprise application store. The concept of the Application Center is similar to the concept of the Apple public App Store or the Android Market, except that it targets only private usage within a company. The Application Center manages mobile applications and you can use the Application Center as part of the development process of an application.

Figure 2.20 MobileFirst platform direct updates accelerates urgent updates

Furthermore, the MobileFirst Application Management feature enables mobile operators and administrators to securely track, search, and control access to users through the mobile applications that are used on their devices, all from the MobileFirst Operations Console. The MobileFirst Platform has a diverse set of capabilities well-suited for taking mobile applications through the Enterprise Lifecycle (Figure 2.21).

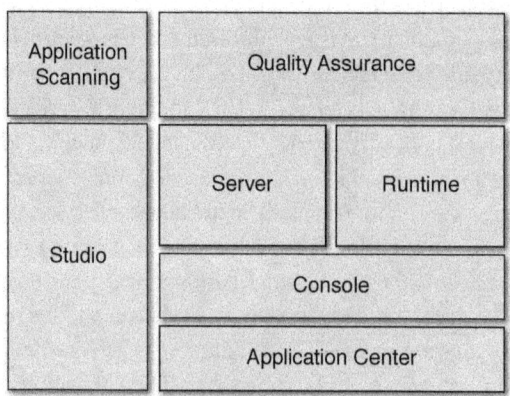

Figure 2.21 MobileFirst platform capabilities for enterprise mobile applications

Summary

In summary, these DevOps Enterprise Mobile Development Lifecycle suggestions are based on a combination of what is being implemented in the market today, and prioritized based on a developer-centric, incremental added value approach for the entire mobile development team. Your actual implementation of capabilities may be different based on your business needs and financial justification and again, for assistance in prioritizing capabilities and aligning them to your business value, consider a DevOps for Mobile Assessment from IBM to offer you such assistance.[29]

---

[29] https://www.ibm.com/developerworks/community/blogs/invisiblethread/entry/diy_with_ibm_s_self_as-sessment_tool_for_devops_practices?lang=en

CHAPTER 3

Design Quality Is Crucial, Make the Investment Up-Front

Application design is typically a subject that is not addressed in software development books. But it is crucial for the success of modern software and no more so than for mobile apps. This chapter covers why design is a critical element of software and discusses some general design principles. We also delve into IBM's specific approach to software design and mobile app design in particular.

Overview

Why Is Design Important?

If you look at popular apps in the market, you will notice some common trends: they have beautifully crafted user interfaces (UIs), positive reviews, and beautiful app icons. In fact, app icons, screenshots, and ratings are the first things people have to view before downloading any app. On the other hand, if users do not enjoy using an app, it is also very easy to delete it and post a terrible review. User experience is not a good-to-have in the mobile market, but more of a norm or an expected attribute. When two apps offer the same functionality, a better design can provide the extra edge. To quote the former IBM CEO, Thomas Watson Junior, "Good design is good business." In this chapter, we will explain some best practices for designing mobile applications that will give you this extra edge and discuss some activities involved in the design process.

Scope of Design in Mobile App Development

When do we say that an app is well designed? Sometimes it is because of the beautiful choice of colors used to create the UI elements, selection of typography, and a clean layout. In other cases, it is because it is really easy to follow and navigate through screens. Even an app that perfectly identifies a problem we face and solves it is considered well-designed. Good design must be usable, desirable, and useful. To create well-designed apps, it is important to understand the different branches of design.

Design Research

The first step is to understand user problems, and then clearly identify how it will help drive business goals. A well-designed app should highlight the core value propositions of the business. This branch of research is called design research. It involves activities like interviewing users and conducting observation studies to understand "why" users do what they do. It also includes methods like surveys, questionnaires, or in-app analytics to quantify "what" most users do.

Information Architecture

When there is a clear understanding of the context of the user, their goals, fears, likes, and dislikes, it is possible to identify different paths users will take to achieve their goals. Once they are using the mobile app, they need to know where they are in the app, where they can go to achieve different goals, and how they can go back. If you think of your app as a house, different screens are like different rooms of the house. Different spaces like rooms, closets, hallways, and nooks help achieve a set of goals. There is a clear way to get from one end of the house to the other that is highlighted in the blueprint of the house. Creators of an app should sketch this "blueprint" that identifies the different screens to achieve user goals, and make it easy to navigate between these screens. Relative priority of user goals should be identified. Common tasks and important goals should be highlighted and should be easier to get to. This branch of design is called information architecture.

Interaction Design

Within each screen, the user interacts with different UI elements to accomplish certain goals. It is important that these elements are intuitive and that they immediately convey their purpose and how one can interact with them. For example, an element like a toggle switch makes it clear that there are two options and at a given instance the user can only select one alternative. Intuitive does not mean that the UI element should only adhere to conventional patterns like "toggle switches" or "buttons" but that its purpose and intended use should be apparent immediately. In fact, new UI elements show up constantly in mobile apps and these styles often uniquely identify the personality of the app and of the people who use them.

Visual Design

People usually identify good design with beautiful looking UI. In fact, it is the typography, color palette, shapes, and grid structure that is immediately visible to the user and conveys to their emotions. This branch is called visual design and it brings information architecture and interaction design together to make an app usable and desirable. Since visual design appeals to the visceral emotions of the audience, it also helps in creating a perception of your business and core values in the minds of the user. Although information architecture defines the hierarchy, it is ultimately, the visual design that makes this clear to the user with the help of correct weights, colors schemes, etc. It also clarifies the interactions and gestures available on a given screen.

Overarching Design Principles and Guidelines

The IBM Design Language is a shared vocabulary for doing great design. In this chapter, we highlight how some of these principles relate to designing for mobile applications. We will specifically talk with reference to the six universal principles.

Discover, Try, and Buy

Before the users start using the app, they would have to discover that it even exists. If this is through a marketing website, then it should clearly highlight what problem the app tries to solve. It is highly likely that the user opens this website on their mobile devices. Hence this website should be designed especially for a small screen size. If your app is a counterpart of a web application, then there are high chances that users will try to open the web application on their browsers on mobile devices. A lot of times, the users will try the web application first, without downloading the app. It is important that the web application has an acceptable experience for smaller screens and advertises that the user can download a mobile app instead (Figure 3.1).

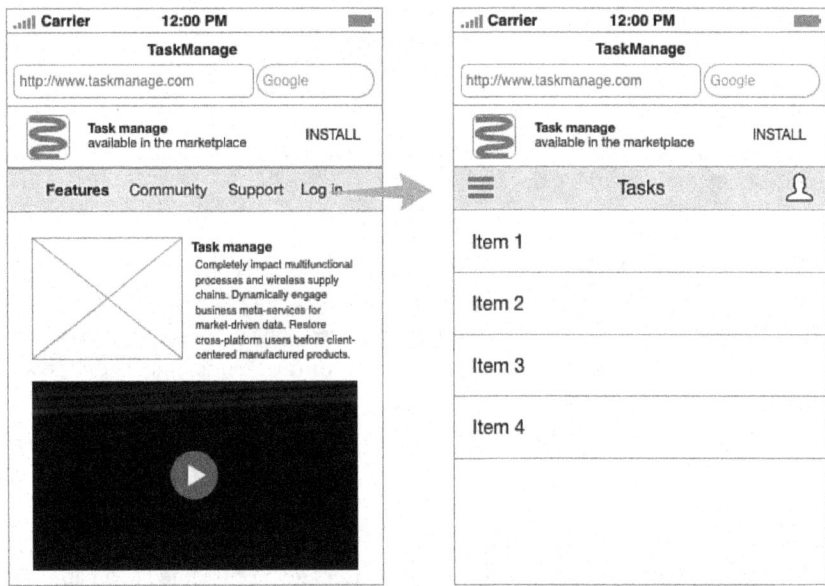

Figure 3.1 The website is optimized for the mobile experience in terms of navigation, size of the buttons, keyboard input, etc. It also reminds the user that an app download is available

App market places are another avenue where people might discover your app. The screen shots should clearly demonstrate how the app solves a set of problems.

Get Started

Every user has a set of goals when they want to use your app. When they open your app for the first time, they need to feel confident that they can achieve these goals (Figure 3.2). Creators of

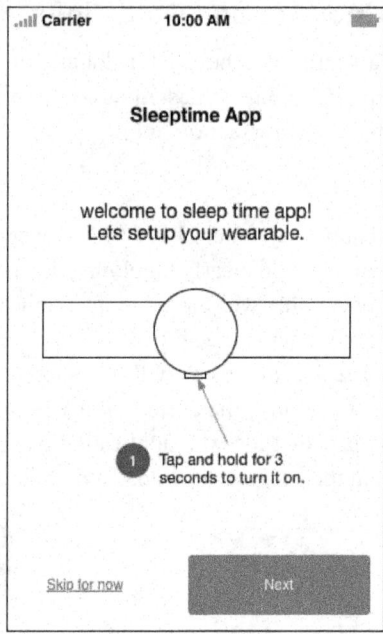

Figure 3.2   An onboarding deck in a "Sleeptime" app when the user opens an app for the first time and has to set up their wearable device

the mobile app should think of such goals while designing the getting started experience. Sometimes this can be achieved by highlighting a clear first step. If there are multiple ways of getting started, then the app should walk the user through a guided tour highlighting available options. A simple on boarding slide deck or a step-by-step demo of different elements on the screen are some ways of accomplishing this (Figure 3.3). The getting started experience might be over a stretch of time. In such cases, the app can track the usage and prompt the user about an unused workflow.

Everyday Use

Although a "getting started experience" helps the new users get accustomed to an app, the design should not hinder the experience of existing users. A well-designed app identifies and reduces repetitive tasks. Some apps provide access to most frequently or recently accessed items. Batch processing a group of items like support for sharing, deleting, or exporting multiple photos at once is another example. The to-do lists app that automatically marks some tasks as important if the user has prefixed an "!!!" (Exclamation mark) is another example. User's location, time of access, current context can also be used to reduce the workflow. Using in-app analytics can help identify such repetitive tasks the users are performing every day (Figure 3.4).

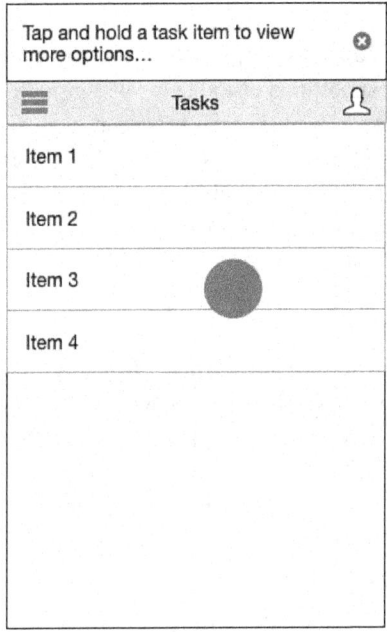

Figure 3.3 A step-by-step demo of the different features and available gestures in the task management app

Figure 3.4 Wireframe of a navigation application: the app provides suggestions from the user's history based on their location and time of the day

Manage and Upgrade

When there is an update to the app, the user should be made aware of the new capabilities. A prompt screen describing the capabilities of the new version with an action to update is a common way apps do this. If the user is not using the app, sending push notifications or emails might be another approach. After upgrading, a "getting started experience" to demonstrate the new capabilities is recommended if there are significant changes. Upgrades should not change the user preferences or their documents, but it is important for the users to know about such changes to maintain the trust with your app.

Leverage and Extend

A mobile app does not exist in isolation but is part of a larger ecosystem of different apps and the mobile operating system installed on the user's device. It is, therefore, important for the app creators to think about the flows between multiple apps. Let us say for example, that a task management app can import notes, rich text documents, or photos into a task item. If so, the user should have the option to export such documents from any app (like a mobile browser or a photo management app) into this task management app. Additionally, such an app should allow exporting a task item into maybe an email app.

Get Support

The mobile app should make it easy for the users to get support that they need. This could be through a well-designed getting started experience, enabling the user in providing feedback to the developers, or supporting a community of users to share expertise and inspire others (Figure 3.5). Apps that have complex workflows also use quick help overlays or a section for short demo videos.

Designing for Enterprise Mobile

The mobile landscape offers exciting challenges for design due to the nature of device and context of use. Since there are so many different form factors in mobile hardware, designing for different screen sizes and resolutions need critical attention. Also mobile by its very definition brings context into the discussion. App developers and designers need to consider different environments, movement, connectivity, attention span, and security among many other attributes.

As our technology further evolves, the mobile landscape encompasses varied definitions such as wearables and Internet of Things. This makes designing for mobile context more complex than just for phones and tablets.

By following a rigorous design methodology, mobile app development teams can ensure user centered design for their apps, thus gaining competitive advantage.

Designing the IBM Way

IBM has recently refocused its journey to be design led. Garnering inspiration from a rich legacy of design innovation and disruption from IBM's past, and at the same time sourcing inspiration from current well known design practices, IBM has developed its own version of user centered design methodology suited to enterprise offerings.

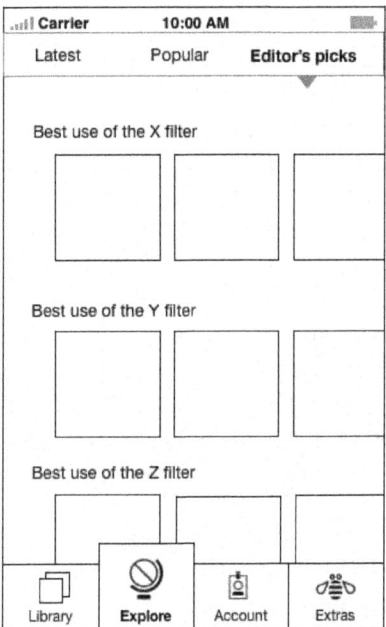

Figure 3.5 Wireframe of a photo editing app that shows "Explore" section where users can share innovative ways of using X, Y, or Z filters and inspire others

Core Practices

The three core tenets of IBM Design Thinking are hills, sponsor users, and playbacks. These address key requirements such as goals, target customers/users, and iterative communication.

Hills

Hills ground your team around important user problems the app hopes to solve, not a list of features and functionalities.

Hills are meant to serve user needs and align larger team efforts towards a deadline. Each individual Hill should be concise, clear, and well scoped so teams have a clear target to work towards.

To ensure each release is feasible, teams are encouraged to have limited number of hills, thus, keeping release cycles agile and focused. Hills are not ambiguous and grandiose as mission statements and at the same time, not a restrictive definition of feature/solution sets.

Example of a good hill is, "My app user can figure out how to use the complete app in first 2 minutes after download."

Sponsor Users

Sponsor Users facilitate participative design by including representatives of your target user throughout the design process. Find sponsor users that represent 80% of your target user.

Having sponsor users identify needs, suggest wants, and evaluate the experience goes a long way towards ensuring applications' success.

Playbacks

Playbacks are the critical communication tool that keeps teams and users aligned throughout the production cycle.

In IBM, playbacks not only serve as milestone checks but also useful tools of storytelling. Playbacks lend themselves as tools that articulate user needs, outline a solution's direction, and describe key attributes that can make a customer go, "wow." Through playbacks, further reviews are initiated, feedback captured, and iteration paths recognized.

Design Thinking

One of the core principles of Design Thinking framework is to ideate and generate out of the box solutions, and then converge around the most desirable pathway. Design Thinking facilitates, lateral concepts, delightful experiences, emotional impact, iterative journeys, and disruptive outcomes among other concepts. IBM uses the following four mental frameworks used in iterative permutations, to root Design Thinking outcomes.

Understand   In the rapidly changing enterprise landscape, the users of apps are also becoming the buyers or in the least have a direct impact on buying decisions. This makes user research a key component in user centered design processes.

Understanding the users enables business outcomes that are profitable. User research also facilitates course correction by solving for the correct problem, not just a desired problem.

Explore   Explore is where the magic happens and art influences the science of Design Thinking. Through exploration processes, designers can ideate, diverge, brainstorm, and empathize around the best solutions to user needs. Exploration serves to benefit from out of the norm and thus provide differentiating, hence competitive outcomes in the market.

Due to the novelty of the space, mobile experiences can still be disruptive in nature and every passing generation of applications seek to push boundaries a little farther. User experience designers are constantly seeking to harness the power of form and context to provide solutions that are more beautiful and better than the ones before.

Prototype   Prototypes enable testing of the product in intermediate stages of the lifecycle. This ensures rapid iteration should there be issues with the outcome. Prototypes ensure we fail early, fail fast, and fail cheap.

For mobile applications, prototypes become significantly important, as they can be tested in context. Core attributes such as motion, touch, location, and interaction can be tested early on for any course correction required.

Evaluate   During Evaluation stages, prototypes of ideas and the applications are run by a representative sample of end users to get first hand feedback. Evaluation seeks honest insights and hence is not synonyms of demos or presentations, but simulation of its actual use environment.

Interestingly, the mobile universe also enables user feedback through app store reviews and ratings. This is one of the best avenues for direct user opinions and rapid iterations. To ensure user satisfaction, app development teams benefit from using such feedback to further enhance the experience.

Some Design Methods

There are many design methods and tools that can be adopted for specific needs. Usually once the app idea is defined and a time line is agreed upon, a suite of design tools and methods can be selected to solve specific needs during the process. The following is only a sneak peek into the wide array of an ever-evolving design arsenal and are commonly used in IBM's design world.

Understand

The following are some of the common design methods used towards seeking rich and in-depth knowledge into users motivations and frustrations.

Ethnography

Ethnography is a commonly used exploratory research method for observing users in their regular environment over a period of time. Such observation outcomes result in understanding context to habits and user patterns.

User Interviews

User interviews are conducted in one-to-one meetings. Researchers talk to users about their needs, wants, and motivations so those responses can be used to influence the application outcome.

Contextual Inquiry

This is a specific form of ethnographic research that combines user interview with work behavior and environment observations. The interviewer conducts such an interview in the natural environment of use. The insights are thus informed by what the users "say" and what they "mean."

Affinity Mapping

Affinity mapping or diagramming is a common tool used to organize and synthesize the research data (Figure 3.6). Each interview point is put on a note and multiple such notes are mapped or grouped to each other based on common patterns of behavior, thought, or desired outcome.

Surveys

To get a quick pulse check on the users and get basic demographic information, a survey is an effective tool. It is used when the application team needs reasonable data in a short amount of time and with few resources. Surveys can be sent out online and users are expected to provide valid data.

Brainstroming Ideas   **Affinity Diagrams**

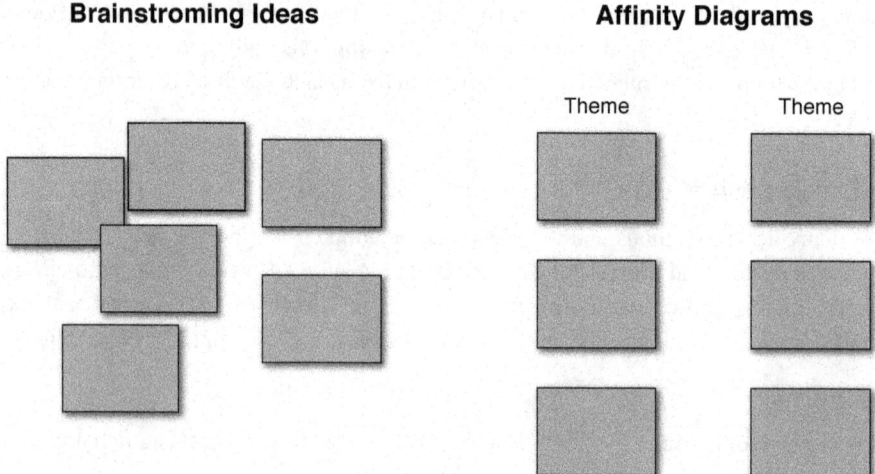

Figure 3.6   Affinity mapping from a divergence of brainstorming ideas

Surveys should be inherently user-friendly. They should not be long winded, demand a lot of cognitive ability from the users, or seek complex decision-making.

Stakeholder Interviews

During exploration phase, it is imperative that application teams get a clear understanding of business strategy, release needs, and key customer requirements. To get this holistic perspective, stakeholder interviews are a valuable tool. These allow for establishing focus towards a common goal.

Competitive Analysis

Competitive analysis is not only useful from a business profitability standpoint but also from user experience vantage point. For user experience, such analyses are carried out around predecided attributes, such as ease of login, user management, community support, etc.

Explore

Systems Mapping

Systems maps serve as tools to document the complex and facilitate visualizing the complete picture. Our users do not operate in isolation and to ensure the applications enable the best user experience, it is important to consider all aspects of the user's world (Figures 3.7 and 3.8). To put it simply, systems maps take a blob of words and phrases that apply to users and make connections, grouping, lead points, etc.

Systems Map

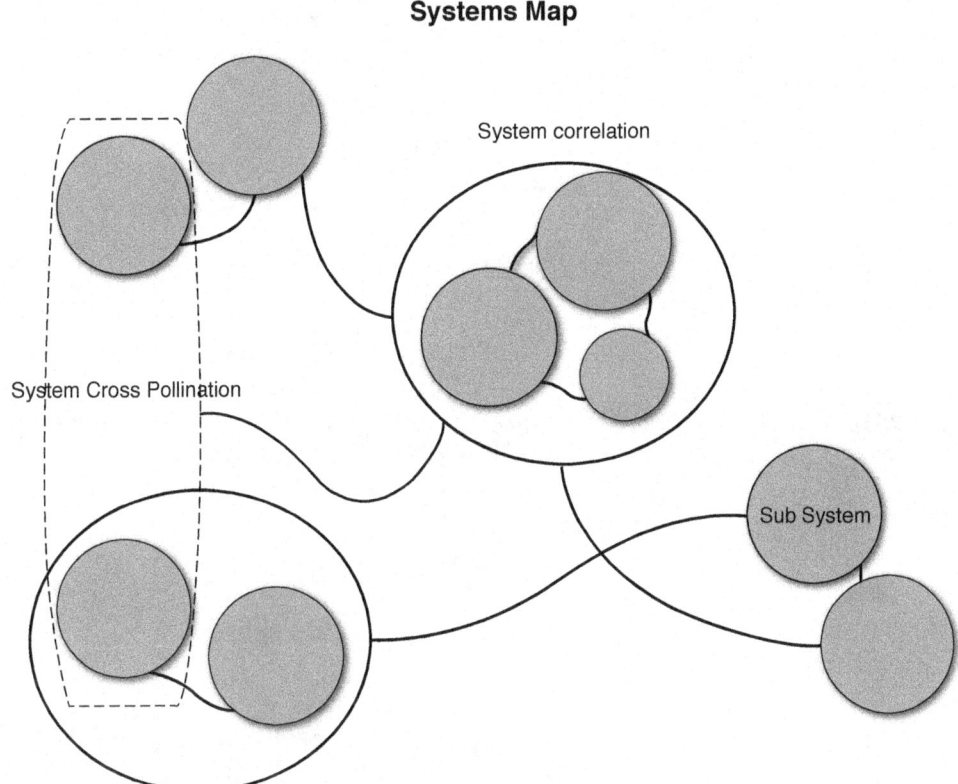

Figure 3.7   Forming relationships and connections through a systems map

Needs Statements

Need statements offer a valuable format to capture user needs and offer valid reasoning. They are also useful to generate new ideas around user needs. Some common formats to capture needs are:

> John wants to pay his credit card on the go so he can avoid going to his bank branch.

> Jill needs a live bus map so she knows when the next bus gets to her stop.

Empathy Mapping

Empathy maps enable the application team to deeply empathize with their user and thus connect emotionally. Such a process helps develop solutions that appeal to users at a visceral level thus creating a loyalty base. Empathy maps templates come in many different forms due to different needs and the one shown in Figure 3.9 is just one example.

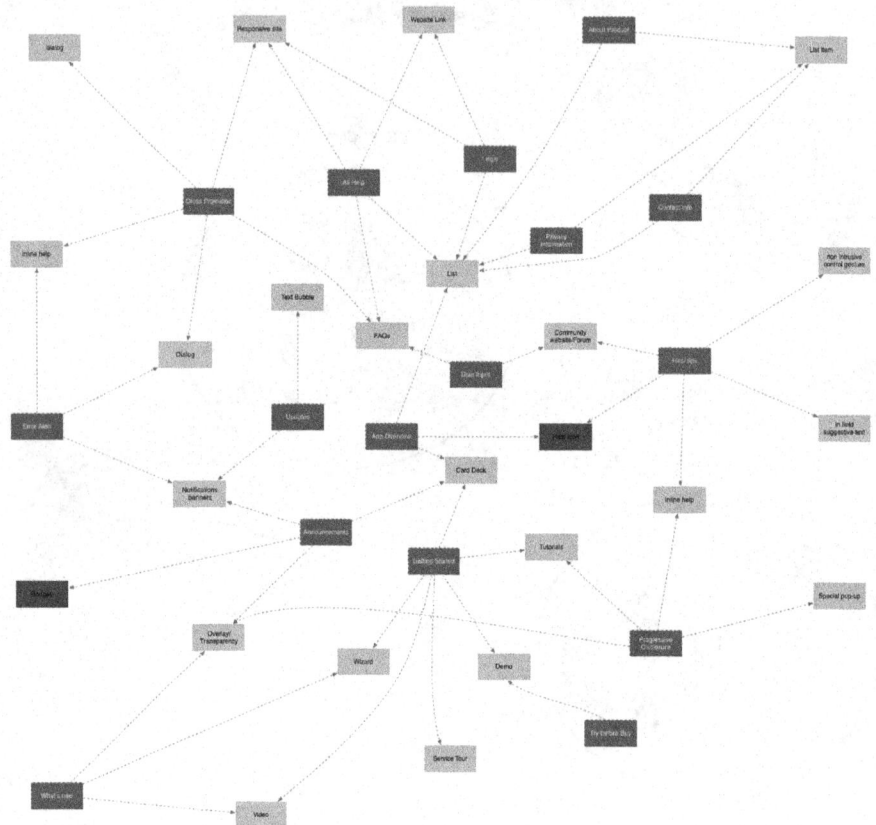

Figure 3.8   Another example of systems mapping

Experience Journey Map

Experience journeys seek to document a user's complete experience from start to finish while trying to solve a particular problem. Journey maps can either document an existing product use or a solution path, a user might use in absence of any application in that space. Experience journey maps capture highs and lows of a user's emotional involvement while also documenting each touch point of significance (Figure 3.10).

Idea Grids

Idea grids are available in numerous formats, and teams can even create their own based on specific attribute needs. IBM uses the one shown in Figure 3.11.

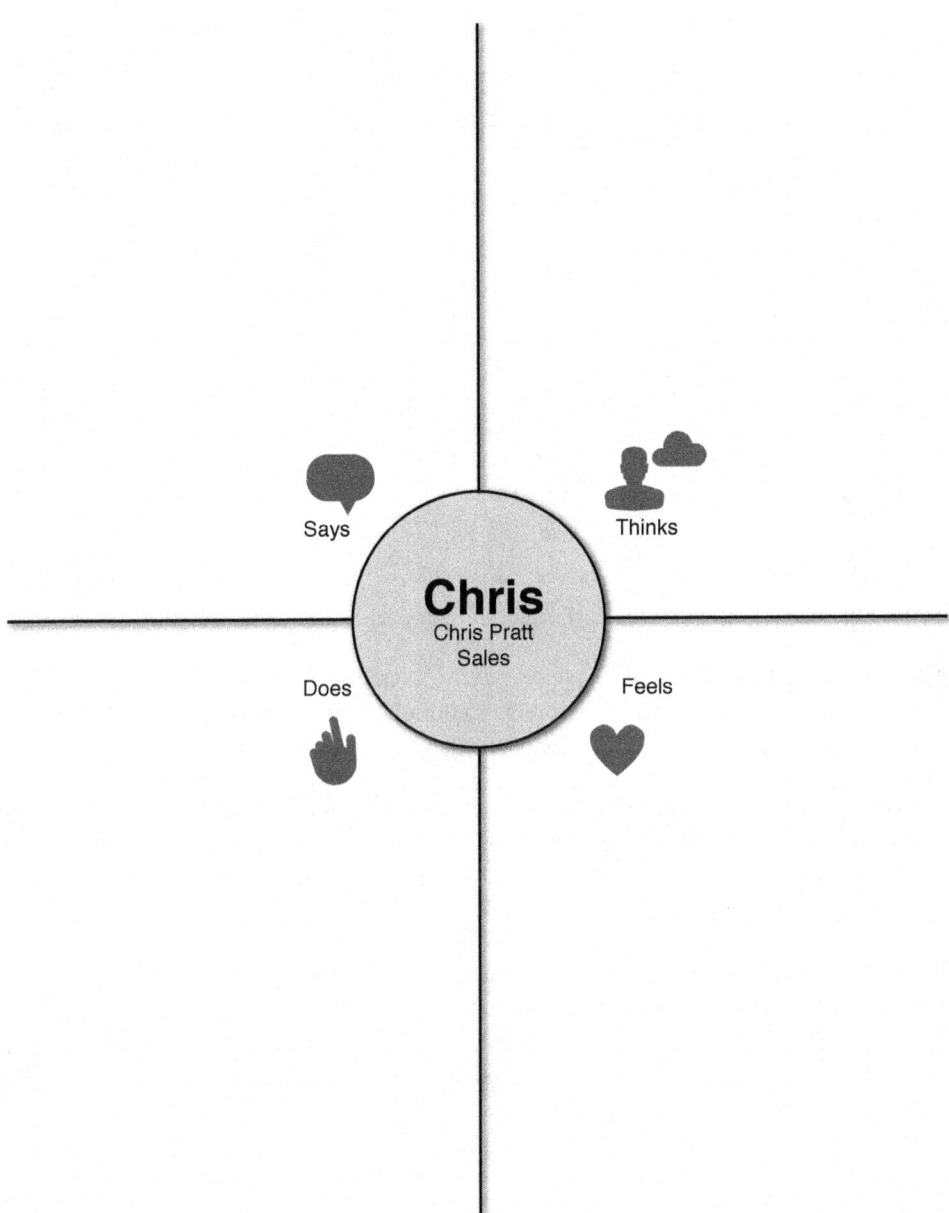

Figure 3.9   Empathy map

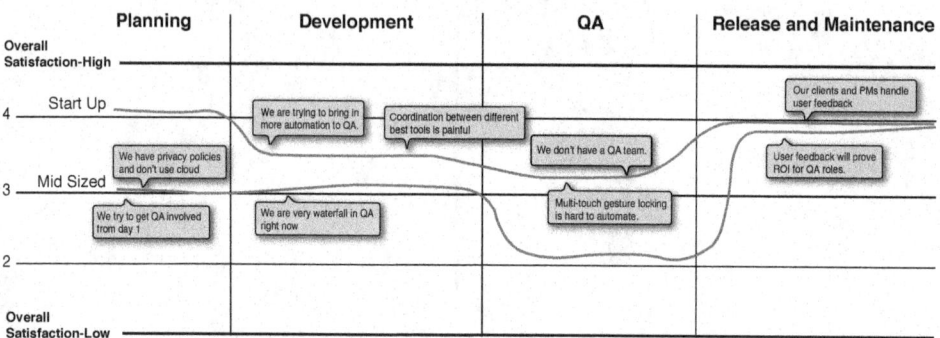

Figure 3.10   As-is experience journey map

Prototype

Low-fidelity Prototyping

There are many different forms of lo-fi prototyping methods. A couple is highlighted as follows.

Paper Prototype   Prototyping on paper is the quickest and cheapest way to test early ideas for success. Not only do users feel safer critiquing a nonfully formed product, but also the advantages of low cost and faster turnaround time cannot be denied (Figure 3.12).

Wireframes   Wireframes allow testing user-interaction flows without any distraction of visual subjectivity. Usually wireframes form an important stage before fully formed visual outputs as interactions and flows are tried and tested out before making the big resource commitment of visuals and end product polish (Figure 3.13).

Hi-Fidelity Prototyping   Hi-fi prototypes mimic finished products very closely in look, feel, and interactions. They are not fully coded products in terms of functionality though. The advantages of developing hi-fi prototypes are that they can be tested with users to get feedback as close to real context as possible before applications are fully developed from both backend and front end perspectives (Figure 3.14).

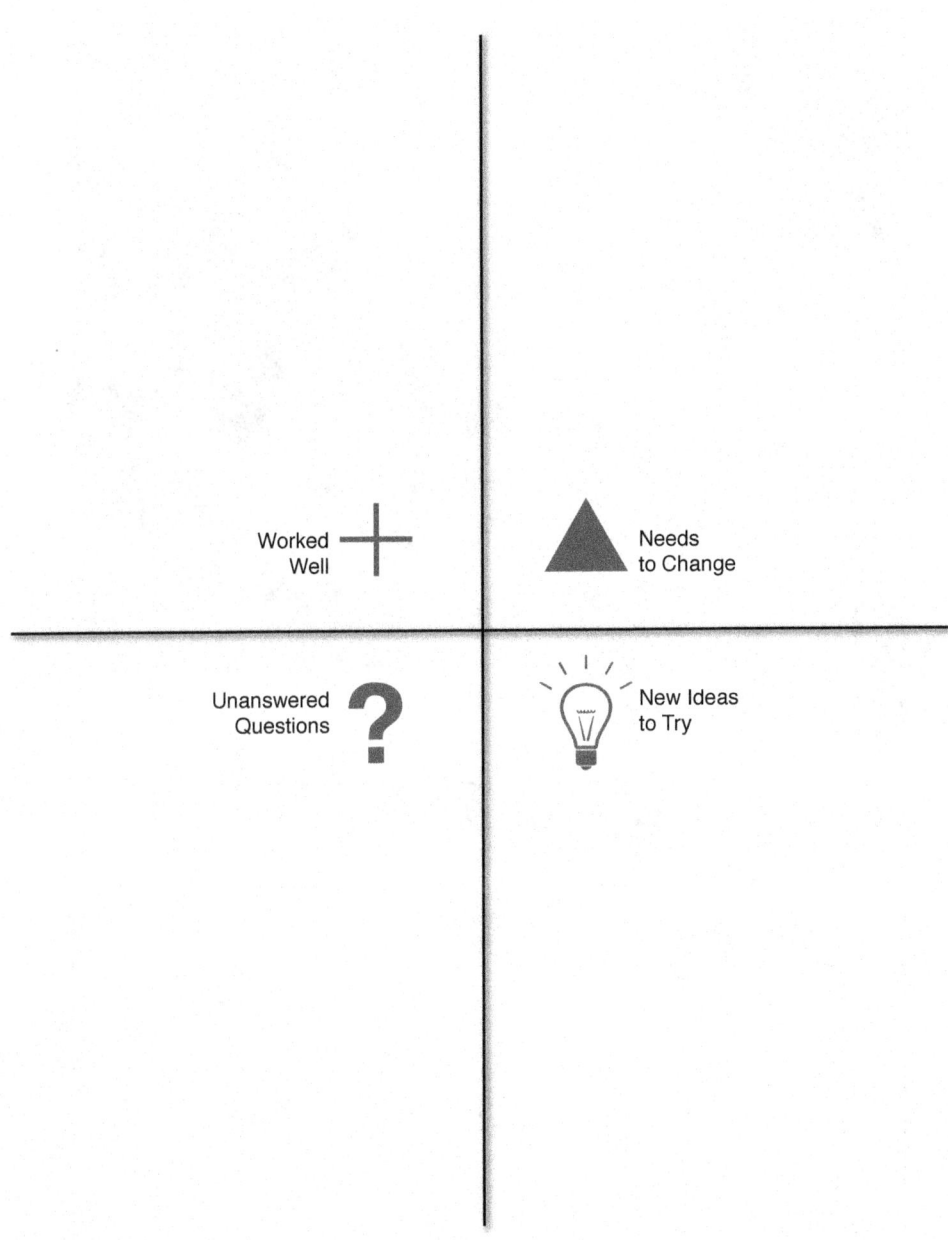

Figure 3.11 Idea grid

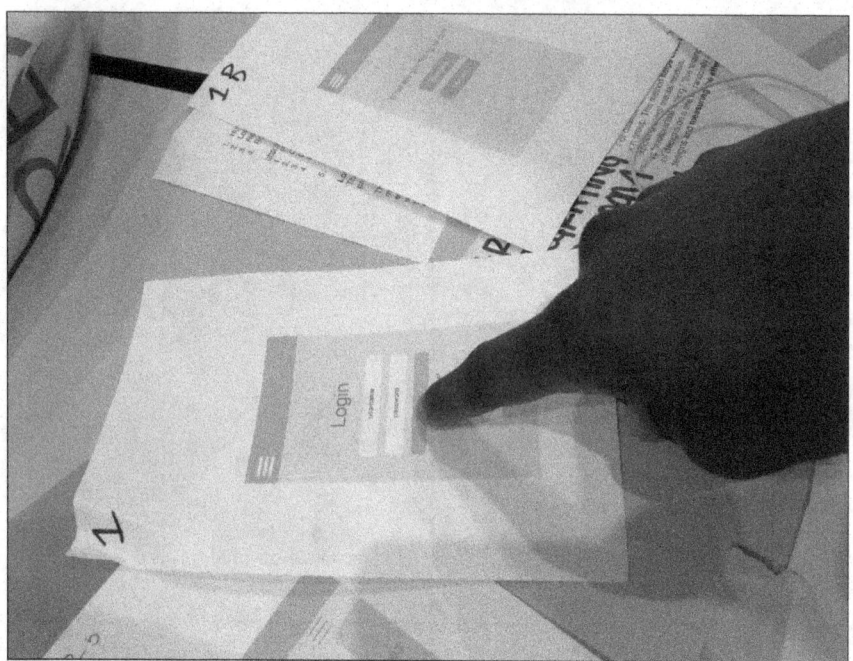

Figure 3.12   Paper prototype user test

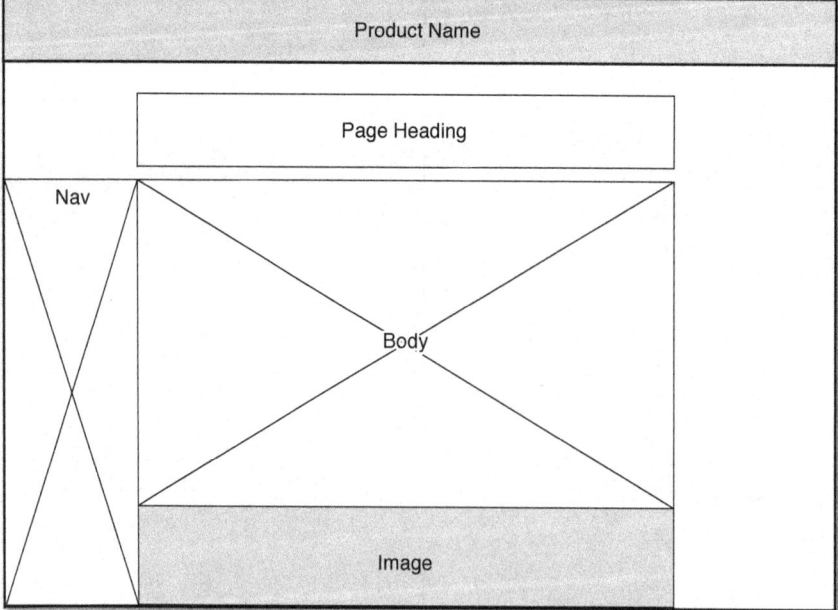

Figure 3.13   Basic wireframe

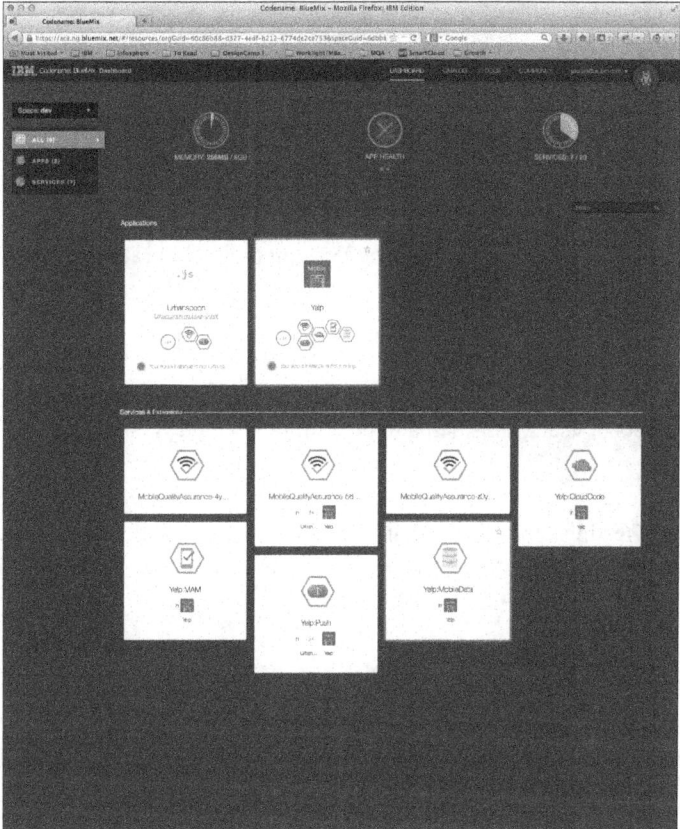

Figure 3.14 Hi-fidelity design artifact

Evaluate

The following methods are commonly used during evaluation of initial ideas, testing fully formed, beta prototypes, and in general getting a sense of users feedback to solutions.

Surveys

Surveys are again, used for evaluative feedback too. Specific questions around likes and dislikes, or ease of use can be asked of users through surveys.

Advantages of a survey are that it can be deployed as close to a user's immediate experience as possible so their short-term memory can be harnessed. Disadvantages include lack of validity as there is no way of knowing who takes the survey and in what context.

Heuristic Evaluation

Heuristic evaluations are inspection tools that test for experience or usability problems in the application. A predefined set of principles or heuristics serve as the measurement criteria for such testing. Common heuristics for mobile applications are privacy, response time, touch areas, consistency, support, etc.

User Testing Interviews

User testing of an application or a product is usually carried out to find out if there are issues with workflow or experience. Application teams can define tasks they would like the user to do with the application and then ideally, leave them alone to mimic real life scenarios. All successful and failed interactions should be documented to iterate the application for intuitive ease of use.

A/B Testing

A/B testing is a method through which two versions of the same application can be release to same or different sets of users. This helps in determining which version fares better with users.

Beta Testing

The closest form of testing a fully formed application with a controlled group of users is to release a closed beta version and gather feedback on how it works in its truest circumstance of use.

Summary

Through this chapter, we introduced you to the importance of user-centered design in the mobile application development process. Design influences the outcome from the very first stages of scope definition, through design research, till the end of the process where the solution takes form through visual design.

Through design principles and guidelines that can be defined by any organization and product teams, user-centered design process can be executed in an organized manner and you can ensure no part of the outcome is unaffected by design.

IBM, through its renewed foray into design, has been influencing enterprise mobility in a significant manner. In this chapter, we shared with you some of the core tenets of IBM's design practice that are uniquely applicable to IBM and enterprise, as also the overall framework of the design methodology.

CHAPTER 4

Mobile Application Development

It was only June 29th, 2007 when the first iPhone came out, but it feels like we have been living with these devices and their apps for much longer. Smartphones have created a new way of living. We are always connected, have instant access to information, instant directions to any location, and much more. The mobile experience is now expected, and enterprises have to adapt and provide these experiences to their customers, employees, and partners. In this chapter, we will discuss Mobile App Development in the enterprise. The enterprise challenge goes beyond the app that runs on the device. The challenges reside in the areas of data, security, analytics, engagement, and so forth. In this chapter, we will focus on the enterprise challenges in mobile. We will focus both on the challenges for both developing the client side app, as well as focusing the core capabilities enterprises need to provide in the cloud.

The Mobile App

When Enterprises began building mobile apps, one of the big areas of concerns was the proliferation of client side technology. Enterprises had become accustomed to standardizing on a platform, such as Java EE. As such, they could contain a skill set around a standard architecture. Mobile devices come with their own application SDK's, as such a proliferation of client side choices have emerged. For Apple, you can write applications using Objective-C or Swift. For Android, you can write applications using Java. Microsoft and Blackberry also have their own SDK's. They all, to some degree, can also support building apps using HTML 5, JavaScript, and CSS, referred to as hybrid development. Right now Apple and Android dominate the major markets, as such, some enterprises can afford to build out two platforms for the best native experience. Figure 4.1 shows the various choices you have for building apps.

In the extreme cases, you can deliver web content through a mobile browser, often called mobile web, or you can build a full native app with the SDK of the device. In between, you have the notion of hybrid based development, where you bundle HTML code within an app. Some hybrid applications are built fully with HTML 5 based technology; others use a mix of Native and HTML 5. Some applications are primarily native with a small amount of web content. Some applications are mostly web and access some native capabilities.

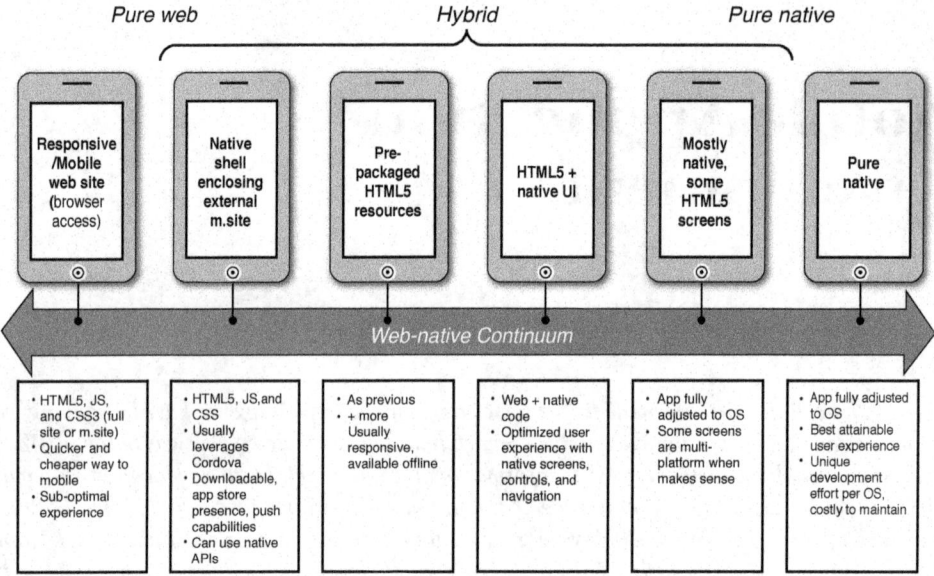

Figure 4.1   The spectrum of mobile app implementation choices

Factors for Choosing

Determining which development approach to choose can be difficult, and it will mostly depend on which requirements you need to fulfill, and how much flexibility you can afford.

> **Business Factors:** There is a clear difference in how much mobile influences develop-ment in different industries. For example, enterprise customers will have very differing requirements than gaming companies. As such, business factors are often a telling gauge for the style of development. Financial industries for example, build out applications that people may constantly use. Requirements include quick access to the latest account information, different visualizations of spending habits, depositing a check with your camera securely, and so forth. As such, a native application with the best experience may be best. Other industries, such as insurance, may not require constant usage, but have key use cases that need to work when accessed, such as filing a claim.

> **B2C or B2E/B2B:** One thing to keep in mind is that mobile plays not only in the B2C space but in the B2E/B2B space as well. The services you provide for your customer as compared to those provided for your employees and partners may have very different characteristics. In addition, while you may be able to control your employee's devices (company owned devices, Device Management Services, etc.), you cannot control your customer's devices. Mobile provides huge opportunities for field-based employees to have instant access to data. Industries like travel, utilities, and insurance provide great opportunities for new mobile patterns for employees.

Target Audience: Understanding your target audience is by far one of the most important factors in selecting a development style. How technical is your audience? Are you trying to reach an audience that is familiar with existing mobile applications, or one for which this may be their first mobile application?

Cost Revenue: Many companies have already realized the financial gain of mobile; however, there are applications you may write that only add to operational cost. For example, providing services to your existing client base may be a matter of survival, but produce no additional direct revenue. Certain internal applications oriented towards your employees may offer little financial benefit as well, but only have intangible benefits, for instance, in improving employee morale.

Frequency of Usage: Some people think about high-use social applications like Facebook when considering applications on a mobile device. People use these applications often and spend a great deal of time on them. However, not all mobile applications, especially in business, adhere to this principle. For example, how often will a person look at his or her insurance policy? How often does an insurance company want you to process a claim? Sometimes, an application may go months without being either used or updated, but you have to make sure that it works when the use is required.

Data Visualization: Mobile devices offer touch capabilities and unique ways to interact with apps. Many business monitoring and reporting capabilities can take advantage of such views. Viewing financial trends, customer trends, etc. are examples of such use cases. However, other data, such as a list of policies, your last five bank transactions, and other data of this sort may require much less visualization.

Multi-Channel Business Context: Sometimes a mobile device is used to do only part of a job. You may start a transaction on a mobile device, and the finish it on a laptop. For example, I might start on insurance claim at the point of an accident by taking a picture, but later provide much more detailed information when I get home. You have to consider how easy is it to move that business context across channels from an integration standpoint when designing your applications.

Good Enough Not to Notice: Usability is a very important factor, proven by Apple through its mobile innovation. The performance of the experience often time plays a role in user perception. However, with the advances in Web Standards and hybrid, sometimes the difference between a native experience and web experience is not noticeable. Technologies like Ionic offer smooth looking transitions between mobile parts of the application, and end users might not be able to tell the difference. However, sometimes you may have advanced graphics and animations where the difference is noticeable. You need to determine how good is good enough, especially if gaining customer reach is important to you.

Existing Skills: Many enterprise customers have invested in developer skills in the area of web development. If customers have a large investment, it may not be advantageous to hire expensive and in-demand native developers, especially if you need to target multiple platforms.

Integrating Across Organizational Boundaries: Many enterprises build applications that interact with other business applications from different groups or different partners. Within Web sites, sometimes this integration is via service calls, sometimes using techniques like embedding web content in an iFrame from somewhere else. Apps introduce some challenges into this style of integration. For example, integrating across two native apps written by partners may be difficult if they did not anticipate integration up front.

Often, different departments within a business host different web applications and only provide links to each other. Thus, from an end user standpoint it might still look like a single site, even though the technological underpinnings are not integrated. However, an enterprise may not be able to provide six or seven unique apps to their customers. For instance, who would want to look for different insurance applications (separately covering auto, health, and life insurance) from the same company?

How Should I Build Applications?

Now, which should you build? The answer is you can build a successful strategy with any of these approaches, and it all depends on several factors. In general, HTML5 hybrid apps actually require much more discipline and architecture to get right.[1] You can be successful with hybrid apps, but you must have best practices defined and architecture in place. Hybrid applications depend very much on the choices of frameworks you use. There are some Mobile Frameworks that focus on providing Mobile Widgets that emulate native look and feel. Many first generation applications use frameworks like jQuery Mobile. Newer frameworks like Ionic have gotten performance fairly close to the native performance on newer devices. Ionic is based on the Angular based framework which gives you a full MVC stack for building maintainable applications. Figure 4.2 shows an example of an Angular architecture.

Having architecture like this is critical for hybrid. In general though although hybrid apps will save you time coding, you may find the cost of tuning your application across a diverse set of Android devices to negate the value in some cases.

Because the iOS and Android are dominating the market, many enterprises are choosing to build native apps. You can have a successful strategy doing native development. Once you learn the frameworks, tools, and languages, you will find that the closed box MVC model can greatly reduce amount of tuning on the client and such. In addition, as vendors release new native patterns, it takes longer for web frameworks to adapt.

The Case for Cloud

The challenge in the enterprise has shifted from client side choice to the ability to provide capabilities quickly, to any channel. As enterprises take their journey from viewing to transacting to collaborating using mobile devices, various patterns are identified and decomposed into a set of capabilities required to meet the end-to-end mobile app lifecycle. There are three stages in the mobile app lifecycle that need to be considered when planning mobile development:

[1] Brown, Kyle, et al. 2014. *Modern Web Development with IBM WebSphere: Developing, Deploying, and Managing Mobile and Multi-Platform Apps*. IBP Press.

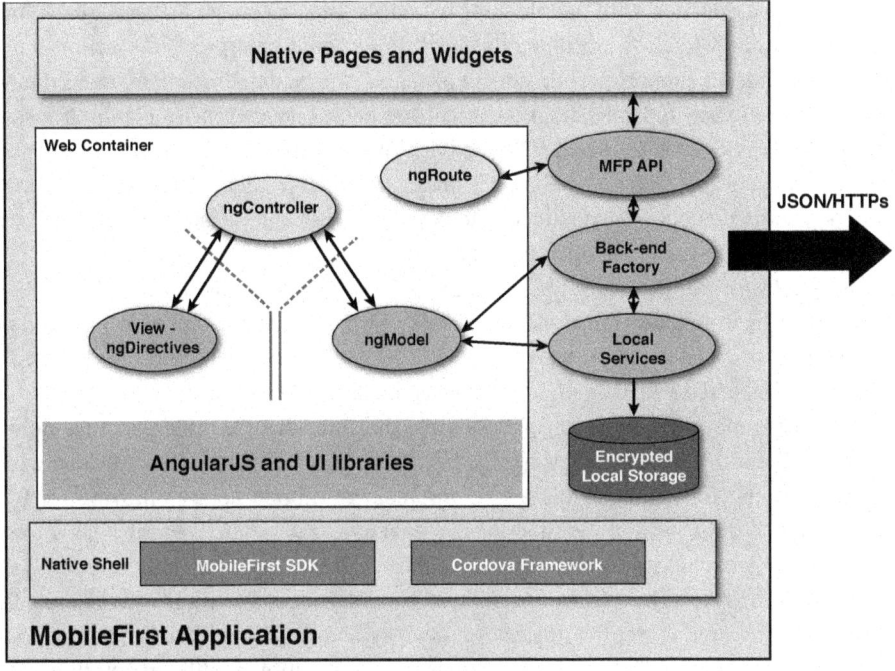

Figure 4.2   Example AngularJS mobile app architecture

1. Developing/deploying a mobile app
2. Running/hosting a mobile solution
3. Mobile app end user

At a high-level, cloud capabilities for mobile support the lifecycle of enterprise mobile applications that are deployed to employee or customer devices and provide managed access to backend business applications and enterprise data sources which support the mobile apps on the devices. These solutions allow companies to leverage emerging mobile technologies to reinvent customer relationships by engaging them anywhere and anytime the context is relevant.

Cloud computing and cloud services are a good match for supporting mobile devices. Mobile apps tend to have time variable usage patterns that are well handled by the scalability and elasticity of cloud computing—increasing and decreasing the backend resources to match the level of requests from the mobile devices. It is also characteristic of mobile apps to make use of serverside data that is unique to the apps. Some of this data is not enterprise data, such as social data (e.g., Twitter or Facebook data), and there are good cloud services available to incorporate such data.

Some data associated with mobile apps is accessed with a frequency and in a volume and format that can be difficult to accommodate with traditional enterprise transaction-based systems. It is common to support mobile apps with one or more databases containing the data for the app. Such databases typically hold copies of the necessary enterprise data in a form suited to serving the mobile apps, such as JSON data held in a NoSQL database. The elastic provision and support of these app specific databases is one of the notable capabilities of cloud computing. Using app specific databases also reduces the need to access enterprise systems and systems of record, along with a reduction in resource requirements.

Another factor influencing mobile app design is the often global nature of app use. Users accessing the apps from many locations put pressure on the infrastructure, with a need to provide "local" endpoints around the globe to avoid latency issues. Cloud computing is well suited to running the same backend services in multiple datacenters around the world.

When developing and deploying a mobile app, it is important to remember that mobile apps typically have a short lifecycle—they change fast to adapt to new devices and business markets. Therefore, planning for agility is another requirement of mobile apps—allowing for frequent, regular updates to the apps and the functionality that supports them. There is a need to support "2 speed IT"—where enterprises manage the systems of record, on premise, enterprise systems at traditional change cycles while allowing applications at the edge or in the cloud to iterate faster— this includes being able to deploy quickly on devices and the mobile backend. Cloud services are good at supporting DevOps, agile development and operations; with the ability to introduce new versions of apps and backend services rapidly through use of automated test and deployment capabilities paired with automated deployments plus monitoring to validate operational quality.

As with earlier major transformative shifts in enterprise technology, a proliferation of implementation options and deployment topologies can make the adoption of mobile capabilities a challenge.

For example, different mobile device platforms, including Apple iOS, Google Android, Microsoft Windows® Phone, and BlackBerry, each come with their own application SDKs. They all support building apps using portable technology; however, native implementations involving a custom app implementation for each device type usually offer the best user experience. This means that organizations need skills and tools to support developing and deploying across all these devices and implementation types.

This proliferation also contributes to challenges with bring-your-own-device (BYOD) in the enterprise. Enterprises need ways to apply corporate policies to devices, which are allowed access to the enterprise network. This includes a means to distribute and update a portfolio of secure custom enterprise mobile applications for employees to use.

Let us focus on the service provider and service consumer in the mobile app lifecycle using cloud computing. The mobile cloud architecture guidance provided by this chapter can help enterprises understand common architectures that have been proven in numerous successful enterprise deployments.

Figure 4.3 shows the overall high-level logical architectural components for hosting a mobile app. It shows how a mobile device, managed by mobile device management, connects to

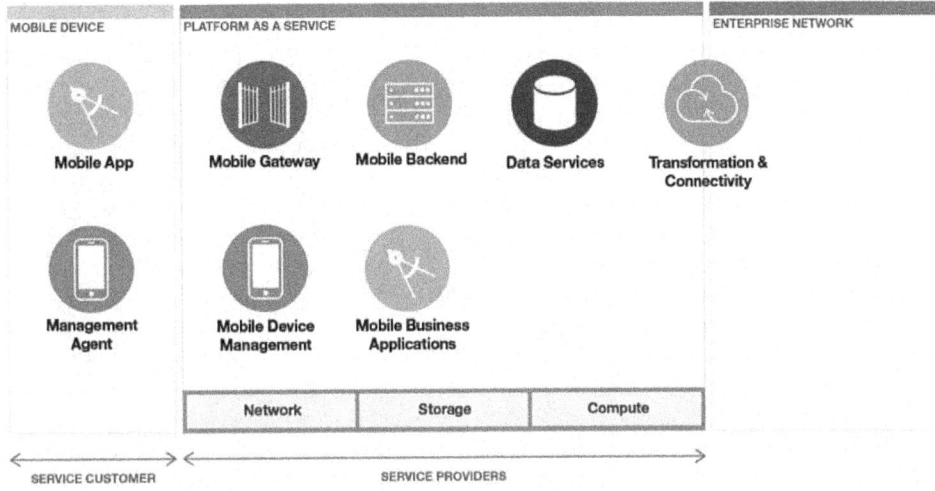

Figure 4.3   High-level architecture for cloud-hosted mobile app

the core cloud components like mobile gateway, mobile backend, mobile business applications, and data services, while transformation and connectivity get relevant data from enterprise systems to show on the device. It shows how a born on the cloud application would use the data services and return the data to the app.

Figure 4.4 illustrates the high-level architecture of a mobile cloud solution. The architecture has four tiers, each containing a subset of the components:

- Mobile device
- Public Network, which connects the device to the mobile cloud services
- Provider cloud service environment, where the various cloud services exist
- Enterprise systems, containing existing enterprise applications, services, and data

Mobile App Architectural Components

Figure 4.4 shows the high-level architecture and its components. This section of the chapter explains the components along with their subcomponents. The subcomponents are also illustrated with a small figure. At the end of these explanations is a mobile architecture diagram with all of the components, subcomponents, and relationships.

Mobile Device Components

- **Mobile App**—Mobile apps are the main vehicle for user engagement with services on mobile devices. Although users can interact with websites through mobile browsers, the use of native mobile apps is the predominant use case.

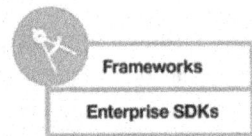

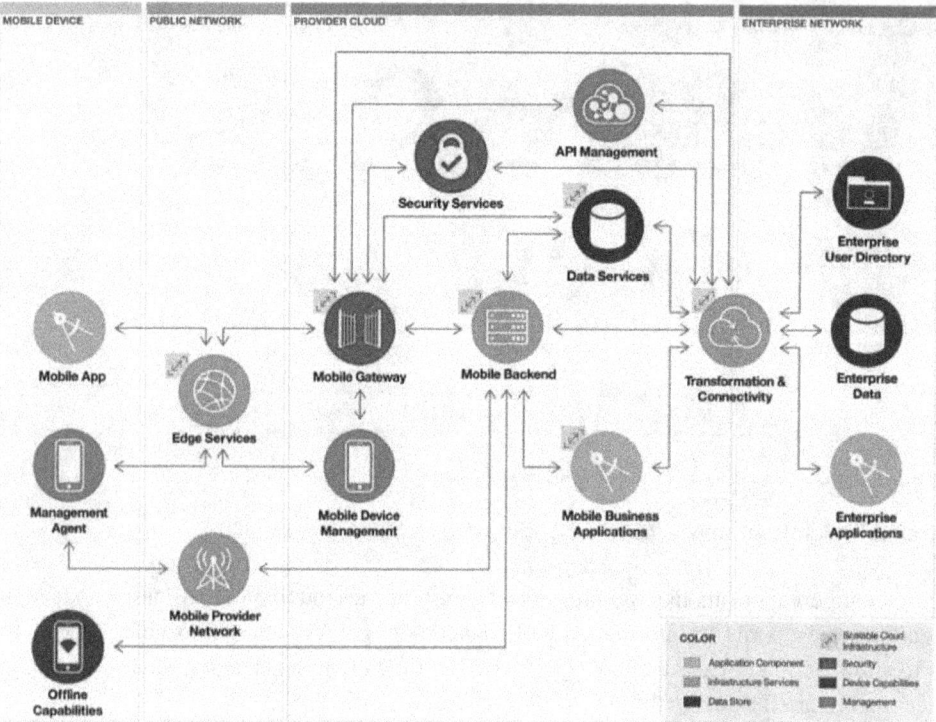

Figure 4.4   A four-tier mobile cloud solution architecture

Mobile apps contain four key components: mobile apps communicate with backend services using APIs, typically based on REST interfaces.

- **Vendor Frameworks**—Provide access to device capabilities and features from the device manufacturer and/or mobile network provider, like Apple Pay, Google Wallet, and Core Data.

- **Enterprise Software Development Kits (SDKs)**—Provide the ability to support communication with mobile backend services through SDK's that are consumable for mobile developers and encapsulates client flows needed to access the backends.

- **Management Agent**—Management agents apply the policies of the enterprise, typically for devices used by employees of the enterprise where the apps deal with sensitive enterprise data. The agent is a part of the SDK that stores, enforces, and manages policies, including security policies, on the device.

- **Offline Capabilities**—Since mobile networks are not always available to the device, it may be the case that a mobile app may use offline capabilities such as an encrypted database to access and store secure data. This component provides the ability to store data securely on devices and sync to the backend when the network is available.

Public Network Components

- **Edge Services**—Edge services include services needed to connect the mobile device and its apps to the right mobile gateway through the Internet using Wi-Fi or mobile provider networks. These include:

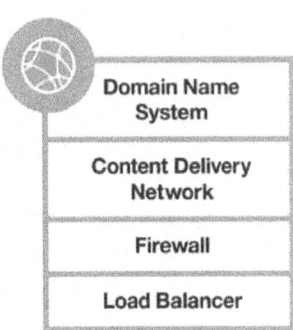

  - **DNS Server:** Resolves the URL for a particular web resource to the TCP-IP address of the system or service which can deliver that resource.

  - **Firewall:** Controls communication access to or from a system—aiming to permit only traffic meeting a set of policies to proceed and blocking any traffic which does not meet the policies. Firewalls can be implemented as separate dedicated hardware, or as a component in other networking hardware such as a load-balancer or router or as integral software to an operating system.

  - **Load Balancers:** Provide distribution of network or application traffic across many resources (such as computers, processors, storage, or network links) to maximize throughput, minimize response time, increase capacity, and increase reliability of applications. Load balancers can balance loads locally and globally. Load balancers should be highly available and without a single point of failure. Load balancers could support any of the components, but support is especially important for the mobile gateway and mobile backend.

  - **Content Delivery Networks (CDN):** Provide geographically distributed systems of servers deployed to minimize the response time for serving resources to geographically distributed users, ensuring that content is highly available and provided to users with minimum latency. Which servers are engaged will depend on server proximity to the user, and where the content is stored or cached.

- **Mobile Provider Network:** Provider of wireless communications that owns or controls all of the elements necessary to sell and deliver services to an end user, including radio spectrum allocation, wireless network infrastructure, back haul infrastructure, billing, customer care, provisioning computer systems and marketing, and repair organizations.

Provider Cloud Service Components

Mobile Gateway—The mobile gateway marks the entry point from a mobile app to the mobile specific services for the solution, typically offering a set of Internet-accessible APIs. The mobile gateway may also use data services and/or the enterprise user directory. A mobile gateway may be implemented by a common gateway across all channels into an API ecosystem. It provides:

- **Authentication/Authorization:** Provides the ability to identify, authenticate, and authorize the user, using a variety of methods and token types. Mobile authentication services provide the ability to handle different token types, like OAuth or OpenID as well as biometric technologies like Voice ID or voice authentication.

- **Policy Enforcement:** Provides the ability to enforce corporate policies during invocations from mobile devices.

- **API/Invocation Analytics:** Provides the ability to capture analytical data of API invocation by a variety of clients (e.g., how often an API is invoked and who is invoking the API).

- **API/Reverse Proxy:** Provides the entry point of an API, usually in a DMZ. API proxy routes an API call to an implementation instance such as an application in the mobile backend.

Mobile Backend—The mobile backend provides runtime services to mobile applications which implement serverside logic, maintain data, and use mobile services. Mobile backend provides an environment to run application logic and the implementation of APIs. Application logic hosted here can communicate with the enterprise network as well as other services and applications outside the service provider. It provides:

- **Application Logic/API Implementation:** Provides the implementation of the business logic being requested by the mobile app via the APIs. The implementation may in turn call on other services to provide required function. A variety of runtimes such as Java or Node JS can be used to code the business logic.

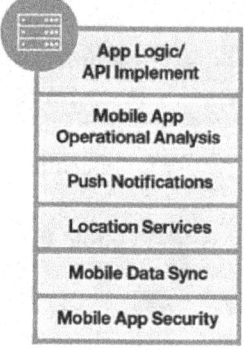

- **Mobile App Operational Analytics:** Provides the ability to do analytics on runtime flows. The mobile backend collects and logs information from mobile apps, such as what client pages were visited, what backend functions were called, what device type called a particular backend, offline storage statistics, and so forth.

- **Push Notifications:** Provide the ability to support subscription and sending of push notifications. Mobile apps allow users to register and receive push notifications while a mobile backend provides APIs for backend logic to push notifications to devices using the mobile provider network.

- **Location Services:** Provide the ability to collect and use location data from mobile apps running on a device.

- **Mobile Data Sync:** Provides the ability to synchronize data on a device and stored in the backend.

- **Mobile App Security:** Provides the ability to do authorization of users to perform app specific tasks; sometimes enterprise security applications are used to provide profile information for this.

Mobile Device Management (MDM)—MDM focuses on managing devices, mostly in Business to Employee (B2E) scenarios. MDM provides services to keep track of enterprise owned devices and also manage devices that connect to corporate networks using management agents on the devices. MDM provides:

- **Enterprise App Distribution:** Provides the ability to host enterprise catalogs and to distribute enterprise applications to mobile devices. If enterprise apps are not deployed to public app stores then enterprise catalogs are needed.

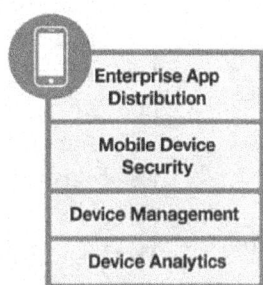

- **Mobile Device Security:** Provides the ability to support enterprise security policies that need to be applied to devices. This includes policies on accessing enterprise networks, password standards, encrypted documents, device wiping, and so forth.

- **Device Management:** Provides the ability for an enterprise to view its organization-wide device usage as well as administrators to be able to add, remove, wipe, and perform actions across all of those devices.

- **Device Analytics:** Provides the ability to capture metrics on the actions performed by employees on devices that can help improve management of devices.

Mobile Business Applications—Mobile business applications represent the enterprise or industry specific capabilities that need to be available to devices that consume mobile services or drive communications with users of devices. These can provide the gateway to enterprise applications and data, and include their own analytics components to track usage. They can include:

- **Proximity Services and Analytics:** Provide analysis and insight into patterns of activity in a physical location to optimize operations or facilitate next best actions. It connects insights from digital activity and physical presence to enable unique engagement with populations and the individual. It also enables contextually relevant mobile communications delivered at the right place at the right time.

- **Campaign Management:** Delivers contextually relevant experiences to connect with customers using mobile apps. This includes using different styles of Push, (Apple/Android) Passbook, Wallet, and SMS solutions. It connects with the mobile backend services and helps to send personalized messages to mobile device users and dynamic sets of individuals based on expressed preferences. This component applies deep analytics to help marketers and app developers understand mobile user behavior, preferences, and usage, thus enabling them to quickly deploy mobile campaigns with relevant offers. It includes the ability to personalize mobile offers in real time, and execute cross-channel marketing campaigns.

- **Business Analytics and Reporting:** Provide complete mobile visibility by capturing user information for mobile websites including both network and client interactions and touch-screen gestures such as pinching, zooming, scrolling, and device rotation. This component can be used to build and manage an early warning system to detect mobile user problems and provide proactive awareness into mobile application failures, usability issues or other obstacles that lead to failed transactions, abandonment, poor app store ratings, and negative feedback. It can also help quantify revenue impact and segmentation by analyzing specific mobile user behaviors or device attributes.

- **Workflow/Rules:** Orchestrates the flow of information at various points in the mobile architecture. A mobile client is integrated and synchronized with mobile business applications, mobile backend, and enterprise systems that are potentially based on different workflow /rules engine.

API Management—API management capabilities advertise the available services endpoints to which the mobile gateway has access. It provides API discovery, catalogs, connection of offered APIs to service implementations and management capabilities, such as API versioning.

- **API Discovery/Documentation:** Provides the ability for mobile developers to find and use APIs securely.
- **Management:** Provides a management view into API usage by mobile apps using information from mobile gateway, backend, etc.

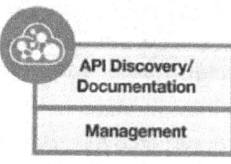

Data Services—Data services enable mobile app data to be stored and accessed. Mobile applications deal with data from many different sources. For example, a user's information exists in enterprise systems, on social networks, and a variety of other sources. Data is often stored in a form suitable for rapid access by mobile apps and sometimes includes (potentially transformed) extracts of enterprise data. Data services can include:

- **Mobile App Data/NoSQL:** Provides the ability to store data in a form that is easily and rapidly consumed by mobile apps.
- **File Repositories:** Provide the ability to store static files, such as PDFs and content. Many mobile applications read files and having a File Store is necessary.
- **Caches:** Provide the ability to cache data for fast access by mobile apps.

Security Services—Security services enable management of access so that only authorized users can securely access mobile cloud services. This component also provides protection of data across mobile devices and cloud services, and enables visibility to have actionable security intelligence across cloud and enterprise environments.

- **Identity and Access Management**: Identifies and authorizes the user providing risk and context based access to mobile and cloud services, including user management, authentication, identity federation, single sign-on, and mobile access management capabilities. These capabilities are leveraged by other components in this architecture—for instance, mobile gateway enforces user authentication and mobile access management, while enterprise secure connectivity enables security services to connect to enterprise security systems like LDAP registries.

- **Data and Application Protection**: Enables protection of enterprise data using a multilevel defense approach across infrastructure, application, and data layers. Application security enables security as part of the development, delivery, and execution of mobile apps, including libraries/tools to secure and scan mobile apps as part of the application development lifecycle. This component helps eliminate security vulnerabilities from mobile apps that access critical data before they are placed into production and deployed. Protecting deployed applications against application threats can be achieved through deploying web application firewalls. Data security capabilities support securing and monitoring access to data in mobile devices, enterprise databases, file shares, document-sharing solutions, and big data environments that may be accessed through the mobile platform, including encrypting data at rest integrated with enterprise key management, secure data in motion through secure connectivity architectures, and data activity monitoring that provides both real time data monitoring as well as vulnerability assessment. Infrastructure security capabilities are enabled by the edge services and the mobile device management components in this architecture.

- **Security Intelligence**: Enables comprehensive visibility and actionable intelligence that can help detect and defend against threats through analysis of events and logs and correlation and detection of high-risk threats which in turn can be integrated with enterprise incident management processes. These same capabilities could also enable with automated regulatory compliance and audit with collection, correlation, and reporting capabilities.

Enterprise Transformation and Connectivity—The enterprise transformation and connectivity component enables secure connection to enterprise systems and the ability to filter, aggregate, or modify data or its format as it moves between mobile components and enterprise systems. Data transformation may be required when the native format of enterprise data is not appropriate for transfer to mobile devices.

- **Enterprise Security Connectivity:** Provides the ability to securely integrate with enterprise data security. Enterprise systems typically hold directories of user identities, such as an LDAP registry, and the ability to access it, authenticate and authorize access to enterprise systems.

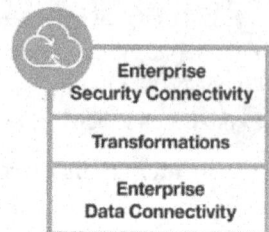

- **Transformation:** Provides the ability to transform data between enterprise systems and mobile components.

- **Enterprise Data Connectivity:** Provides the ability for mobile components to connect securely to enterprise data. Examples include VPN and gateway tunnels.

Enterprise Network Components

- **Enterprise User Directory**—Provides storage for and access to user information to support authentication, authorization, or profile data.

- **Enterprise Data**—One or more systems of record, for example, transactional data or data warehouses that represent the existing data in the enterprise.

- **Enterprise Applications**—Applications that run enterprise business processes and logic within existing enterprise systems.

Complete Picture

Figure 4.5 shows the complete picture for the **Cloud Customer Mobile Architecture** with all of the components, subcomponents, and their relationships. Each of these components has been explained with a subcomponents figure in the previous section.

Mobile App Flow

Figure 4.6 illustrates the flow of a typical use case for mobile banking. The mobile user installs the mobile app on their device, and then uses it to deposit a check to an account by taking a picture of the signed check from the mobile device. The bank also offers services to subscribe for text or email notification when certain events occur, such as an account falling below a minimum balance or possible fraud alerts. This scenario has three different flows:

1. Mobile app installation flow number 1 in blue.

2. Check deposit flow numbers 2–8 in yellow.

3. Push notification flow numbers 9–10 in green.

- The banking customer installs the mobile app onto their device after browsing a public application marketplace such as Google Play or the Apple App Store. In an enterprise usage scenario, the company—using their own enterprise app store and corporate mobile device manager—may instead push the application to the device over the public

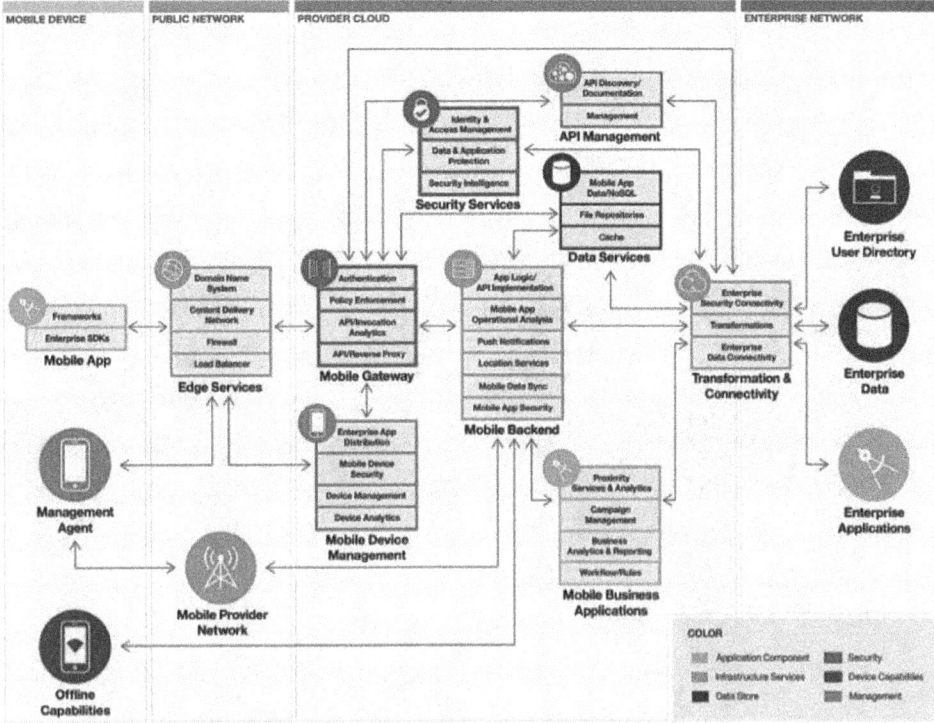

Figure 4.5 A complete view of the mobile cloud solution

network. As part of the installation process, the user can opt-in for location aware services and sign up for push notifications on account balance changes or fraud alerts, for example.

- The customer then uses the mobile app to deposit a check by taking a picture with the camera built into the mobile device. The user logs into the app (which will communicate with mobile gateway for authentication) and then sends the "deposit check" request to the bank with the check image. The user interaction is logged for understanding customer behavior and for understanding operational efficiency.

- This service can be located using DNS, load balancers, and other public network boundary components collectively known as edge services. For all transitions from the mobile app on a device to the mobile gateway through the public networks, which can be wireless or mobile networks, the mobile app sends requests using a URL resolved by a DNS to an IP address. The IP address may be the IP address of a CDN server, load-balancer, firewall, or proxy service in front of the mobile gateway. The CDN server determines if the requested content is in the CDN storage network. If the CDN server cannot satisfy the request the request is sent to the firewall. The firewall evaluates the request and

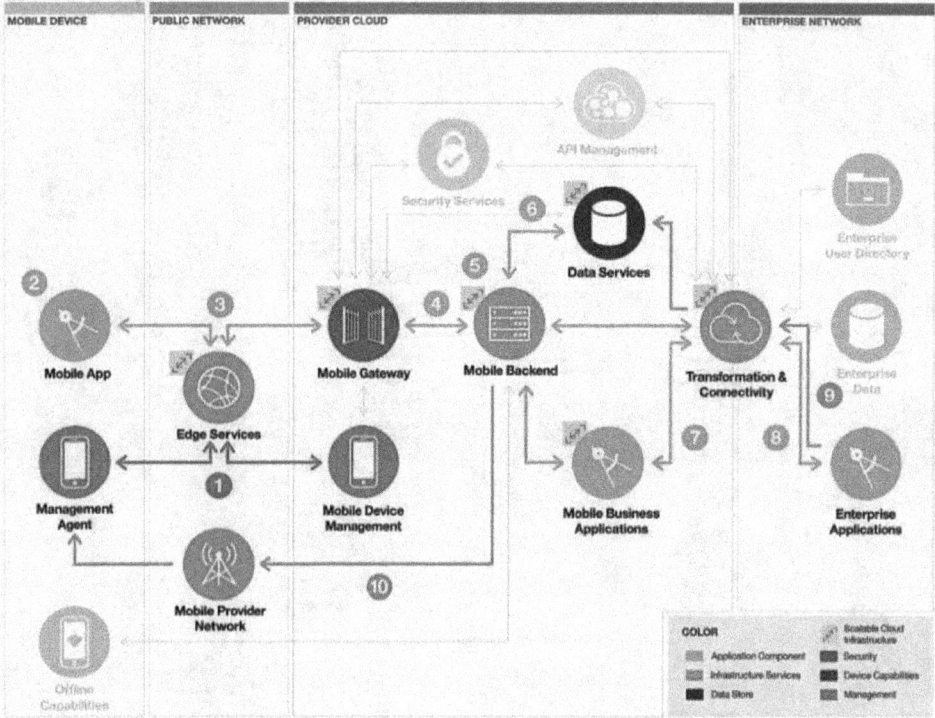

Figure 4.6   Mobile cloud solution architecture applied to a typical banking app

allows the request to continue forward to the mobile gateway if it meets the firewall rules.

- The mobile gateway receives the deposit request and checks security rules for access to the "deposit check" service and uses an API service lookup to direct the request to the right service implementation in the mobile backend. The security check validates credentials and authorization of actions and passes a successfully validated user request on to the mobile backend. It then logs the activity for analytical purposes.

- The mobile backend executes the "deposit check" business logic to store the check image and send the check information to backend processes and systems to deposit the check in the customer's account. The service retrieves information through the transformation and connectivity components that enforce enterprise application security and ensure the account is valid. The mobile backend provides location services and manages subscription services for push notifications. It logs the activity for analytics usage. The mobile backend uses the workflow/rules service to start the deposit check process flow.

- Data services may be used to speed up response time—for example, the account balance could be stored in a NoSQL database and check images may be cached in the file repository. The "deposit check" business logic now stores the account, image of the check, and the deposit amount using the data services APIs. The information is logged for analytics.

- The mobile business application workflow sends the deposit check transaction through the transformation and connectivity components that enforce enterprise application security rules and grant access. The process flow uses different services to check the validity of the check, store the image of the check in the enterprise document repository, and deposit the check using the core banking application. The process execution step is logged for analytics.

- The enterprise account application stores the image of the check in the enterprise database for tracking purposes and applies the deposited amount to the customer's account in the core banking application. Control returns to the mobile backend. When the mobile backend "deposit check" service application completes its tasks, the resulting content is delivered through the mobile gateway (which logs information for analytics again) and the public network to the mobile app.

- Sometime later, once the amount is added in the enterprise application, a request is sent through the transformation and connectivity components which use the data services API to update the account with amount deposited and the balance. This information is now cached in data services for expedited access and to save resources by reducing accesses to enterprise systems.

- Depositing of the check information in data services invokes the push notification service in the mobile backend to send an alert to the customer that the check was successfully deposited. The mobile backend manages subscription services to determine and send the alert via a push notification on the public network to the device. The push notification service takes care of connecting to and using the right mobile provider network.

The customer receives the notification that their deposit has been accepted and continues to interact with the banking application. The context of app usage is recorded for analysis by the bank to ensure ongoing excellent customer service.

Independently, the analytics being collected can be used for a variety of business purposes, including campaign management, fraud detection, and business presence needs. For example, as the analytics determines that this was a large deposit that can change the customer status to an elite customer, it sends information about elite customer service to the customer through push notification. Alternatively, locations service showing lots of activity from a certain geography may cause the bank to add an ATM locally or to engage them in local investment opportunities.

Mobile App Deployment Considerations

Mobile capabilities can be deployed in a number of different ways. One dimension is whether to deploy the mobile services using an Infrastructure as a Service (IaaS) cloud service or whether to use mobile services provided by a Platform as a Service (PaaS) cloud service. A second dimension is whether the mobile services use a private cloud deployment model, a public cloud model, or some form of hybrid cloud model.

Using IaaS cloud services means that the cloud service supplies basic resources such as compute nodes and data storage capabilities—it is the customer's responsibility to install the required software for each of the necessary service components and to configure them to work as a cohesive whole. Some of the components can be bought as off-the-shelf software, while others (such as the mobile backend applications) must be purchased or developed by the customer.

PaaS cloud services typically provide many if not all of the mobile service components as a set of cloud services. The customer selects and configures the services they need—and develops code for custom components such as the mobile backend applications. The deployment of the underlying resources is largely automated by the cloud service provider, with minimal effort from the customer.

For public cloud deployment, the components are instantiated in a shared datacenter of the cloud service provider. For private cloud deployment, the components are instantiated either on-premises within the enterprise or within an isolated environment in the datacenter of a cloud service provider. For hybrid cloud deployment, there is an element of choice of where to locate each component, either in a public cloud environment or on-premises. It is typically the case that public cloud deployment is likely to be lower cost than private cloud deployment, but may have security risks that cannot be accepted by the customer.

In each case, the actual deployment topology used is driven by business factors with the choice typically governed by security and performance considerations. Security and data protection considerations will depend on the nature of the data associated with the mobile apps and the business impact of risks such as unavailability of data or of data breaches. Generally, the greater the sensitivity of the data, the more likely it is that on-premises private cloud deployment is used. However, even if some components are deployed on-premises, it is not necessary that all components are deployed in this way.

For example, an enterprise can choose to deploy the mobile backend components in a globally distributed shared public cloud while keeping the mobile business application services and data within private data centers to meet performance objectives while at the same time ensuring proper security and protection for sensitive data.

The DNS and CDN usually live in the public Internet—these are typically purchased as services from a suitable provider.

For IaaS cloud services:

- The firewall and load balancer are deployed in the cloud service. Many cloud service providers have firewall services available. The load balancer can be run on one or more servers within the cloud service.

- The mobile backend components and the mobile business application services and data components are deployed onto virtual machines, containers, or bare metal nodes provisioned in the IaaS environment. Networks are configured to allow traffic through the mobile gateway, to the mobile business application and API networks, and to the mobile provider networks. Components can be collocated or installed onto separate virtual servers as desired.

- Significant consideration is required concerning the number of instances deployed for each component. For resilience and redundancy, it is advisable to have at least two instances of each component, preferably in geographically separated data centers. To take advantage of the scalability and elasticity of the cloud services, it is also necessary to increase and decrease the number of instances of a component dynamically according to the work load placed on it. This requires monitoring and management components and also requires appropriate load balancing in place for the component.

- The transformation and connectivity component spans between the cloud computing environment and enterprise system and consideration must be given to how it is structured—is it largely within the enterprise network, or does it mostly exist in the cloud computing environment? Performance and security are key factors influencing the design.

- For any components that involve data storage, such as the data services, considerable thought needs to be given to the number and the location of copies of the data. Replication and backup are necessary design points, as is consideration of the number of compute instances allocated to the components reading and writing the data, and also to the consistency model applied to the data.

For PaaS cloud services:

- The firewall and load balancer are typically part of the cloud service.

- The mobile backend components and the mobile business application services and data components are typically provided by the PaaS itself, only requiring allocation and configuration by the customer. Custom code must be developed for the application components, but this is typically deployed into runtime environments provisioned by the PaaS.

- Data storage components are part of the PaaS and it is typical for these to be provided with options for replication and backup—in the best case with customer control over the locations used to store the data.

- The transformation and connectivity component may be supplied by the PaaS, but it typically requires at least the installation of some connectivity code within the enterprise network.

- For a PaaS, scalability and elasticity are usually built in, although often requiring configuration—for example, establishing a set of policies for when to increase and when to decrease the allocation of resources. Similarly, it is common for a PaaS to support load balancing of replicated components, often done transparently when multiple instances are allocated.

Regardless of where components are deployed—public, private or hybrid—lifecycle, operations, and governance requirements need to be considered and addressed. Where components are deployed will strongly affect how management and governance are done. Private deployments may be able to use existing enterprise management and governance tools if they have access to the cloud infrastructure. Lifecycle operations (instantiate, initiate, and terminate) for components instantiated outside the enterprise need to be agreed on and supported by the cloud service provider for public, hybrid, and externally hosted private deployments. In all cases, the key is automation—as much as possible should be completely automated and manual interventions should be reduced to a minimum.

Similarly, operational monitoring and management interfaces for gathering metrics, checking SLAs, status, notifications, and negotiating changes in capacity for these public components will need to be obtained and support for them should be added appropriately to existing management tools. This may include integrating data, information, tools, and processes from multiple sources into common interfaces, reports, automation, etc. for efficient and scalable operations.

Governance and compliance processes will need to accommodate the change in control and risk over externally hosted components. Optimally, lifecycle management solutions should integrate across deployment models and provide a common, integrated context that enables management of release, change, security, SLAs, problem diagnosis, etc. in a complex, dynamic, and potentially unreliable environment.

Summary

This chapter focused on what to consider when building mobile apps, including the style of app you want to build. More importantly, it focused on enterprise challenges for mobile and how enterprises are turning to mobile capabilities in the cloud to fulfill them.

CHAPTER 5

Mobile Enterprise— Beyond the Mobile End-Point

Enterprise backend "Systems of Record" are a rich source of services and data that can greatly differentiate a mobile application from its competitors. Often times connecting to the System of Record is a necessity. This chapter discusses opportunities and challenges when enterprises build mobile applications that tap into the power of backend systems.

Building Mobile Apps Powered by Enterprise Backend

There is a saying on the web: "the best mobile apps are connected." To provide the most engaging user experiences, mobile applications often reach into the services backend. These services may provide the mobile application the needed intelligence about the user, or use enterprise logic to make real-time decisions. This principle of creating immersive user experiences through connecting the mobile app to a sophisticated backend is also applicable to enterprise mobile applications.

Enterprise mobile applications are often backed by IT services owned by the enterprise or third party services providers. Often the main mission of these mobile apps is to drive business processes that have been implemented and deployed in existing IT systems in the enterprise. This is especially true for employee-facing apps.

In today's age, mobile devices are increasingly becoming an important platform where employees carry out their work activities. In some lines of work, mobile devices have always been a necessity. For example, in the mining industry, asset management and task scheduling for a coal mine are often done on tablet devices because the work needs to be carried out deep in the ground without the convenience of a desktop computer connected to the network. Almost all modern companies have started to build a comprehensive mobile solution for their employees in their various lines of work, as an important alternative to the desktop or laptop computers.

Many of the early pioneers in the enterprise mobility front rely on custom-built mobile devices that are preloaded with the applications. These applications operate in a fully customized and tightly controlled environment. This has changed significantly today with practices like Bring-your-own-device, also known as BYOD, where employers allow work-related activities to be carried out on the employees' own personal devices. In a BYOD practice, it is critically important that the apps installed for work-related activities and the data that flow through these apps can coexist peacefully and safely with the personal apps and data on the same device.

These special circumstances that enterprise mobile applications operate under require certain aspects to be given special considerations. For instance, secured access is of utmost importance. From the same device the user may be living their personal life in 1 minute, and carry out business transactions in the next. Modern mobile operating systems, like iOS 7 and later, allow Virtual Private Network (VPN) to be configured on a per-app basis. This means personal applications can use the open Internet connections while enterprise applications installed from the corporate app stores use the secure tunnels provided by the corporate VPN.

Data protection is another aspect of security that companies should dedicate IT resources to, so that sensitive or confidential data do not remain on the mobile device after the business work is done, or "spill over" (for instance, through copy-and-paste) to nonconfidential communication channels like personal emails.

Enterprise mobile apps also use data stored in enterprise backend systems, or System of Records. To make the data accessible from mobile devices on public cellular networks as well as company Local Area Networks, special infrastructures such as a secure mobile gateway and data caching to enable fast access are required. Considerations should also be given to the data and service application programming interface (API) design to best cater to the different dynamics of mobile access patterns compared to tradition desk-bound accesses from computers. The average time span of mobile user sessions is much shorter and often only deal with a small portion of the data set related to a narrowly focused task.

According to a study by Boston Technology Corporation, in the next 5 years, integration of enterprise backend systems to mobile apps will represent 20% of the total IT spending on integration.[1] In that same study, 60% analyst and 60% solution providers agree that integration with legacy systems is the main obstacle to mobile implementation. No doubt, integration is essential to a successful mobile enterprise strategy, yet at the same time there are many challenges IT architects would have to face.

Connecting the Mobile App with Enterprise IT Services and Data

Most enterprises have built a sophisticated system of backends over time that carry out core business operations, manage enterprise resources, and increase employee productivity. Accessing these systems from mobile devices typically involve going through an integration layer that establishes security contexts and transform data from the System of Records to fit the mobile form factor, which usually means reducing the amount of data to just what the mobile applications need. In most cases, the integration layer are implemented as Representational State Transfer (ReSTful) services, which use JavaScript Object Notation (JSON) as the data interchange format because it is easy to parse and lighter weight than some alternative formats, such as eXtensible Markup Language (XML).

---

[1] Boston Technology Corporation, 2014. *All About Mobile Integration—Stats, Facts and Methods.*

All mobile applications should strive to provide a good user experience. However, depending on the kind of services and data that the applications need, what is happening under the cover for invoking services and obtaining and manipulating data can be very different from one application to another. Using the backend services and data as a way to categorize mobile applications, there are three kinds:

Thoroughbred Cloud-Based—The enterprise backend systems that these mobile applications connect to are not locked inside enterprise IT infrastructures, but are instead deployed in a public cloud. As an example, many companies use salesforce.com for their Customer Relationship Management or build custom business applications on force.com. Since these applications are deployed in salesforce.com, there are no special corporate firewalls to go through in order to access the services provided by these applications, or the data managed by them. A mobile application can connect to APIs in salesforce.com fairly easily. Most of the cloud platforms or business application platforms that exist in a public cloud support token-based authorization, such as OAuth, which makes it pretty straightforward to build mobile applications to log in to the system and gain access to the services and data, without requiring an elaborate security infrastructure.

Traditional Corporate IT—Many productivity-enhancing, employee-facing mobile applications fall into this category, where the backends are deployed behind corporate firewalls and are accessed either from the company intranet or through VPN. It is relatively simple to build a mobile application that can assume existing connectivity to the corporate network. What is more challenging is when a mobile application needs to access the corporate IT systems from the public Internet, which is typically the case with consumer-facing mobile applications for e-commerce. A lot more security infrastructure is needed to ensure only the authorized parties are able to access the services and data is protected from attacks. Fortunately, mobile applications can build on top of the existing infrastructure that have already been set up to enable web applications to conduct online e-commerce.

Mixture of Cloud-Based and On-premise Backends—A lot of mobile applications build on top of the System of Records while giving birth to a new System of Engagements. Data in the System of Engagements are not typically highly classified as those in the System of Records that reflect either trade secrets or a business's core competency. As a result, a public cloud system is the ideal place to store this new kind of data, since it is much easier to access and cheaper to maintain (the cloud platform provider does it for you). Many mobile applications need to work with both System of Records and System of Engagements. This means they must integrate with a public cloud system and are able to securely access corporate IT systems at the same time.

Depending on the types of deployment of the backends used by a mobile application, as described above, and the types of deployment of the mobile application itself, there need to be different strategies to ensure data flows smoothly and securely from the backend to the mobile edges. In the following sections, we will go into more details on the three tiers of architecture involved in creating an enterprise mobile application solution: the backends, the integration layers, and the mobile application platform itself.

Types of IT Backends to Integrate from Mobile Apps

From the vantage point of the mobile applications, not all IT backend systems are created equal. During the planning phase of a mobile application development project, IT architects should make an inventory of the different backend systems that the mobile application will directly interact with. It rarely is the case that these systems are ready to be accessed from mobile applications. Depending on the nature of the system and the specific use case requirements, different integration strategies need to be employed.

The first category is the "Enterprise Information Systems." These systems usually run the core business operations and hold the mission-critical data that define a company's bottom line:

- IBM Custom Information Control System (CICS®)
- IBM Information Management System (IMS™)
- Customer Relationship Management (CRM)
- Enterprise Resource Planning (ERP)
- Relational Database

These mission-critical systems make up the "System of Records." In most enterprises, they are maintained and securely guarded by the company IT. To gain access to them often require layers of approvals. Typical data structures returned from these systems are complex and requires transformation by the integration layer. Good news is that the latest versions of these systems have all acquired ReSTful interfaces to allow much easier integration with consuming applications. For example, IBM CICS Transaction Server V4.2 and later, can install an additional Feature Pack for Mobile Extensions to acquire the ability to use JSON RPC, or ReSTful services to reach the COBOL, C/C++ or PL/I programs from mobile applications. SAP Netweaver offers interoperability with web and mobile applications since its first release in 2004.

Another important factor to consider in building integrations with these high-valued backend systems is the cost to access. Due to licensing terms typical of these commercial systems, cost is usually proportional to the volume of access. Since mobile apps tend to generate volatile usage patterns due to its convenience (it is not difficult to imagine a customer waiting at the airport constantly brings out the smartphone and checks the status of an order), directly accessing the backend system from each request sent from the mobile app is not cost-efficient. A caching strategy is needed to lighten the load on the backend system and better handle the mobile usage patterns.

The second category is systems for internal business processes such as office automation or process management. They are mainly used to boost employee productivity and improve process efficiency.

- Employee Self-Service Portal
- Business Process Management
- Enterprise Document Management

While these systems do not directly impact the company's revenue, they are critical to reduce operation expense and improve productivity. To achieve this, it is important to provide a continuous experience between the original platform's web user interface (UI) and the mobile app. With these mobile apps, providing a consistent UI design between the new mobile frontend and the platform's existing web channels will increase the usability of the apps to the employees who have become used to the existing usage paradigms.

A popular approach to support a continuous experience from the desktop to mobile channels, or "multi-channel access pattern," from a single platform is by employing responsive web designs. Responsive web design is a technique that uses CSS media queries to load different CSS rules under different browser form factors (width, orientation, etc.) to achieve optimal layout of the HTML content. Details of responsive web designs are beyond the scope of this book, but there is tons of information on this popular technique both on the Internet and in printed materials.

Many employee productivity softwares support multi-channel access by employing responsive web design techniques. For example, IBM's WebSphere Portal has a number of mobile-enabled themes that renders individual portlets as full-screen pages when the portal page containing these portlets is loaded in a mobile browser.

Often times though, simply using responsive web design is not sufficient if the mobile channel has a high standard on the quality of user experiences. What you can do with responsive web design is limited to manipulating layout of HTML content blocks and show/hide based on the detected form factors. But the mobile applications have unique user experience patterns that are fundamentally different than desktop web applications. For example, to display a list of records, a mobile application typically uses a master-details pattern like the following (Figure 5.1):

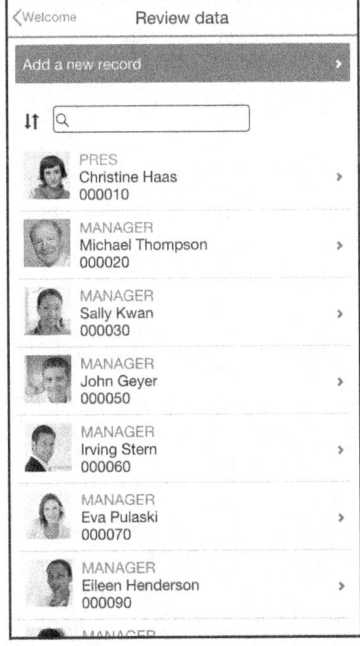
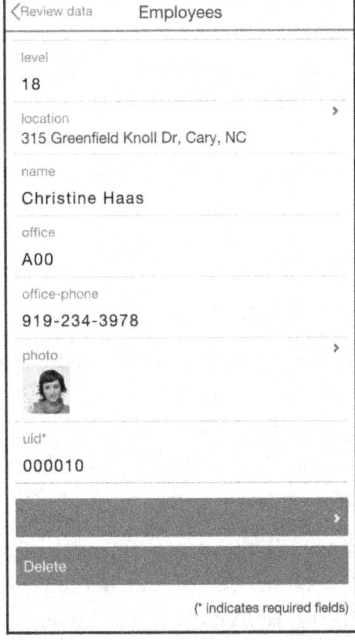

Figure 5.1 Mobile design uses a master list and details view pattern for smaller screens

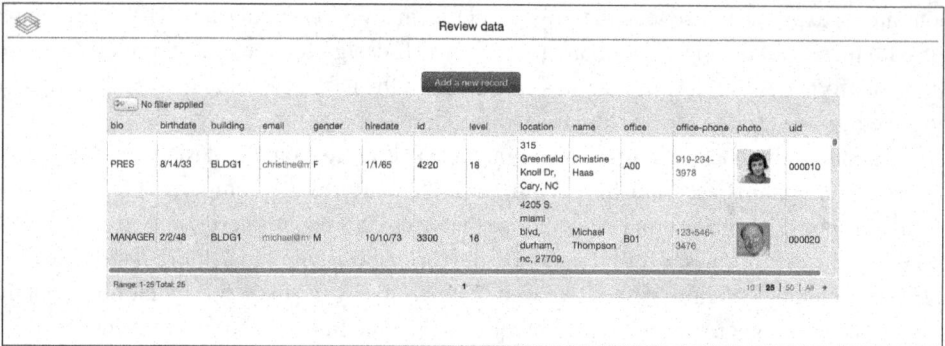

Figure 5.2   Desktop design uses a tabular pattern for faster access to information

But in a desktop browser, due to the bigger screen estate, a grid is usually the best UI to allow fast access to the information (Figure 5.2).

To achieve optimal rendering for both desktop and mobile channels, as shown in Figures 5.1 and 5.2, different widget systems need to be employed. Many JavaScript UI widget frameworks provide a widget set for the desktop rendering and a separate widget set for rendering on mobile devices. Examples include jQuery UI (desktop) versus jQuery mobile, dijit (desktop) versus dojo mobile, and ExtJS (desktop) versus Sencha Touch (mobile). This often implies that the UI will contain separate implementations for desktop browsers and mobile access. Indeed, many web sites will redirect the mobile browser to a mobile-specific implementation that is built with UI widgets designed specifically for optimal user experiences on mobile phones.

Another interesting characteristic of these internal employee-facing systems is that they often have tight coupling between the UI layer and the system API. This is an unfortunate result of the legacy software architectures from the Web 1.0 era, which still exist in the latest versions of many products in this category. As a result, building the corresponding mobile apps against these existing systems often require significant integration efforts and may involve modifying the existing backend system toward a more service-oriented architecture.

Type of API Protocols

Mobile applications integrate with backend systems through APIs. In the last couple of years, most companies have built a layer of web APIs around their core IT systems as part of the Service Oriented Architecture. That is the good news. The bad news, however, is that the web APIs do not have a common standard. Some of the more prominent protocols used in today's backend systems for integration include:

> **Simple Object Access Protocol (SOAP)**—Heavy weight, when compared to the other protocols, has its own technology and specification stack and does not rely on other protocols. SOAP is a transport-agnostic specification and can be implemented on top of HTTP or other protocols like Java Message Service (JMS). SOAP is used widely in machine-to-machine interactions.

Representational State Transfer (ReST or REST)—Used most widely in UI-to-service integration, and increasingly used to replace SOAP as a more lightweight solution for machine-to-machine integration.

Open Data Protocol (OData)—A REST protocol that uses the AtomPub protocol to deliver the requested data. The specification also includes a description format for the service API and data models. OData is used widely in products from Microsoft and SAP. IBM WebSphere eXtreme Scale also supports OData in its REST interfaces.

eXtensible Markup Language (XML)—Still used as the wire protocol in many back-end system APIs for its elaborate data type definition system.

JSON Based Remote Procedure Call (JSON-RPC)—As the name suggests, it is a procedural API that is built on remote functional calls. Unlike REST, RPC style API creates tight coupling between the consumer and the provider. Because of this, more and more JSON_RPC APIs are being converted to REST to allow improved flexibility in implementing consumer clients and service providers.

ReSTful services are preferred for integrating mobile applications with backend systems. It takes full advantage of the HTTP architecture, which is the *de facto* transport layer for essentially all mobile applications, and does not bring along additional semantics as does SOAP, which would complicate the application architecture. In addition, JSON is preferred over XML as the wire format of the data because of its lightweight. This is because when data is used in the UI layer, they often have static binding to the UI controls which gives them semantic meaning without needing elaborate descriptions using an XML schema. OData is a rich set of specifications that follows the REST style. As a result, consuming OData services does not require a huge set of client libraries and can be easily added to a mobile application.

Security Integration

As with web applications, mobile applications must establish a security context with the backend systems before access to the services and data is granted. Security integration between a mobile application and the backend systems can be achieved either through a client-side approach, or a server-side approach. In the client-side approach, the mobile application uses a client Software Development Kit (SDK) provided by the target backend system to interact with the backend directly over the network. In the server-side approach, a server that supports the mobile application interacts with the target backend systems on behalf of the mobile application.

Let us look at the client-side approach first. The mobile application authenticates with the target backend directly from the mobile device, either by logging in through a native login form provided by the target backend SDK, or by using an already authenticated account associated with the device. Many cloud-based backend systems support OAuth for authentication and authorization, which allows the mobile application to redirect the login flow to an OAuth provider like Google or Facebook for the user to establish a security identity through the OAuth token issued by the provider, which then can be verified by the backend systems through the corresponding OAuth provider. Other authentication and authorization mechanisms used by enterprise backend systems include basic authentication and client certificates. Basic authentication is regarded as less secure in most circumstances because it requires the user ID and password to be sent in clear texts in the request payload (although the request packet itself would typically be encrypted and sent over HTTPS). Using X.509 client certificates is more secure but it requires an elaborate Public Key Infrastructure (PKI) where client certificates that uniquely identify a user can be issued and delivered to the client mobile application through a secure process. On the other hand, the PKI can be part of the Mobile Device Management services, which will be discussed in a later section of this chapter.

Mobile operating systems like Android, iOS, and Windows Phone all allow devices to be configured with an account to be used for such purposes like automatic photo backup, email synching, or receiving messages. Some authentication providers provide mobile client SDK capabilities that use these device-identifying accounts, which have already been authenticated, to automatically establish the user identity. This makes for an improved user experience because repeated typing of user ID and password is avoided.

The main advantage of using a client-side approach for security integration with a backend system is that user identity is established and saved on the mobile device, in the form of security tokens or session cookies, so that the server supporting the mobile application does not need to keep track of the user credentials or maintain the user session with the backend on behalf of the application, which lightens the server memory consumption.

Security integration can also be accomplished with a server-side approach. First of all, almost all enterprise mobile applications have a client side, which is the application code that runs on the mobile devices, and a server side, which runs on a server platform or in a cloud. The server side provides security, data access and backend integration to support the mobile application running on the devices. Often times the server code is deployed in a mobile platform, which is sometimes called Mobile Enterprise Application Platform (MEAP) for on-premise deployment, or Mobile Backend as a Service (MBaaS) when deployed in a cloud infrastructure.

Mobile application's server-side component would need to manage the security credentials that are used to authenticate to the target backend systems. This means when connecting to the backend services or data API, the server component will use a user account or set of user accounts that are predetermined at development time or be configured by the application administrator. This is different than the client-side approach where the user of the mobile application establishes the identity through explicit login. The server component typically saves the security credentials in an encrypted storage that is guarded by administrator privilege. When backend services are requested from the mobile application, it will perform automatic login using the stored credentials or present a client certificate that has previously been issued. The client identity that is established this way is completely separate from the user of the mobile application, in other words, the identity of the mobile application's user does not flow through to the backend system invocations.

Finally, let us discuss Single Sign-On (SSO). There are multiple parties involved in the usage scenarios of a typical enterprise mobile application: a mobile user signs in to a mobile application, which has a database or data service for the System of Engagements, the mobile application needs to invoke services from one or more backend systems that may each use their own security provider. Without an SSO solution, the mobile application users would need to sign in multiple times whenever a different backend system is invoked. An SSO solution creates a federated identity system by using a single directory server or security provider. More complex SSO supports multiple security providers by trust-associating multiple security providers and mapping identities among the different providers.

A flow diagram of a mobile user signing in with an SSO based on a directory server, using IBM Security Access Manager and IBM Worklight is illustrated in Figure 5.3. IBM Security Access Manager provides authentication and authorization to mobile application users and devices, federated SSO, and identity mediation across different cloud service providers and web services. IBM Worklight is the MEAP where mobile application's server-side components are deployed. Mobile applications built with IBM Worklight SDKs communicate with the IBM Worklight server for secured data access and services invocations.

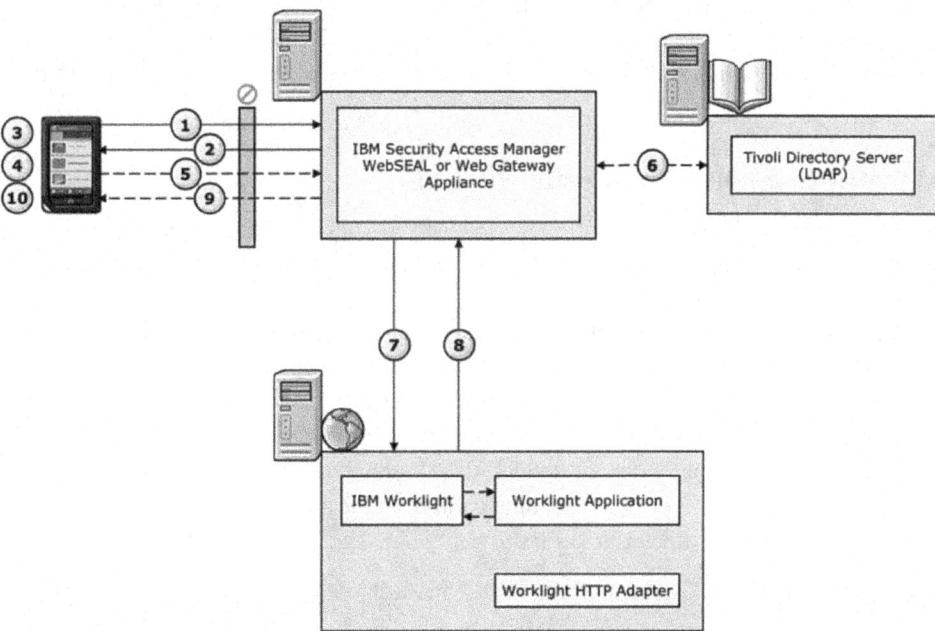

Figure 5.3   Security architecture for mobile applications using IBM Worklight

The client connects to Worklight server (e.g., /flights).

1. WebSEAL, acting as the reverse proxy, caches the /flights request and responds to the client with a 200 OK status response and the WebSEAL login form.

2. The client detects the login form as a custom response, and displays the login form on the device to the user.

3. The user enters their user name and password, and clicks Login.

4. The client performs a POST operation to WebSEAL, providing the login form data.

5. WebSEAL authenticates the user.

6. If successful, WebSEAL forwards the original /flights request to the Worklight server and the configured authentication data, either the HTTP header or Lightweight Third Party Authentication (LTPA) token.

7. The Worklight server, acting as the client, retrieves the authentication data, sets the user data for the realm, and returns the successful 200 OK response to WebSEAL.

8. WebSEAL, acting as the reverse proxy, applies any path or host name filtering, and returns the response to the client.

9. The client continues to run with subsequent requests through WebSEAL (e.g., /flights).

More backend systems can be added to the SSO system as illustrated above, provided that they use the same user registry in the Lightweight Directory Access Protocol (LDAP) server. In the SSO setup, the mobile user's identity flows to all the different systems including the mobile backend, which is the IBM Worklight server in the example above, and the backend systems.

IBM DataPower XG45 Security Gateway

IBM WebSphere DataPower appliances are designed to support the implementation of enterprise solutions by introducing security layers, providing application integration capabilities, and enhancing overall integration performance.

The main advantage of WebSphere DataPower appliances is that they are easy to integrate into a network infrastructure, where they provide a software-independent configuration and simplified functionality. In addition, in development environments, some appliance virtual machines can be used to accomplish the same basic functions of the real appliances.

When used as the security layer provider, IBM WebSphere DataPower appliances can execute many critical security tasks by off-loading them from the application server (Table 5.1).

Key security-related capabilities of IBM WebSphere DataPower appliances are as follows:

- Unloading HTTPS session handling from the web server

- Acting as web application and XML firewall

- Providing authorization, authentication, and auditing (known as AAA) in a single mechanism

- Implementing an enterprise SSO function through the use of LTPA tokens

In terms of application integration and performance enhancement, IBM WebSphere DataPower appliances provide the following functions:

- Unloading XML, XSLT, and XPATH processing from the application server to the DataPower appliance and performing data transformations with better response times

Table 5.1 IBM DataPower XG45 Security Gateway Features

Feature	Description
Web application firewall and gateway	Protects against XSS, SQL injection, XML vulnerabilities.
	Offers extended security functions beyond those of an XML firewall. The extended functions include web service access control (AAA), XML encryption and digital signature, WS-Security, and content-based routing.
	Supports JSON schema validation and JSON payload protection.
	Supports HTTP header (including cookie) signature and encryption.
XML and JSON denial-of-service protection	Validates incoming requests and documents malformed and malicious traffic (e.g., use the schema validation to make sure the message is valid).
	Gain access to valuable postattack forensics.
	Controls the low-byte XML and JSON messages that can bypass your traditional perimeter protection.
	Deploys service level monitor (SLM), service level agreement (SLA) to regulate the incoming messages.
Field-level message security	Selectively shares information of entire messages or of individual XML fields.
Access control for web services	Enables secure access to web services-based applications for your internal and external clients.
Fine-grained authorization	Offers fine-grained authorization that interrogates individual requests to determine whether they can be allowed through.

- Acting as an enterprise service bus (ESB) to provide integration between different architecture layers
- Accelerating data conversion and application integration
- Providing support for transmission of specialized business-to-business message traffic between partners

When deployed in a public-facing, or Business-to-consumer, application, the XG45 appliance (or virtual appliance) is typically used in the DMZ as a reverse proxy, ensuring authorized access from the mobile applications to the internal enterprise systems. In the follow example, the DataPower appliance secures the mobile application's server-side component that is deployed behind the corporate firewall, which in turn integrates with other enterprise backend systems. In Figure 5.4, "Worklight Server" is an example of a MEAP where the server-side component of a mobile application runs and communicates with the mobile application running on mobile devices.

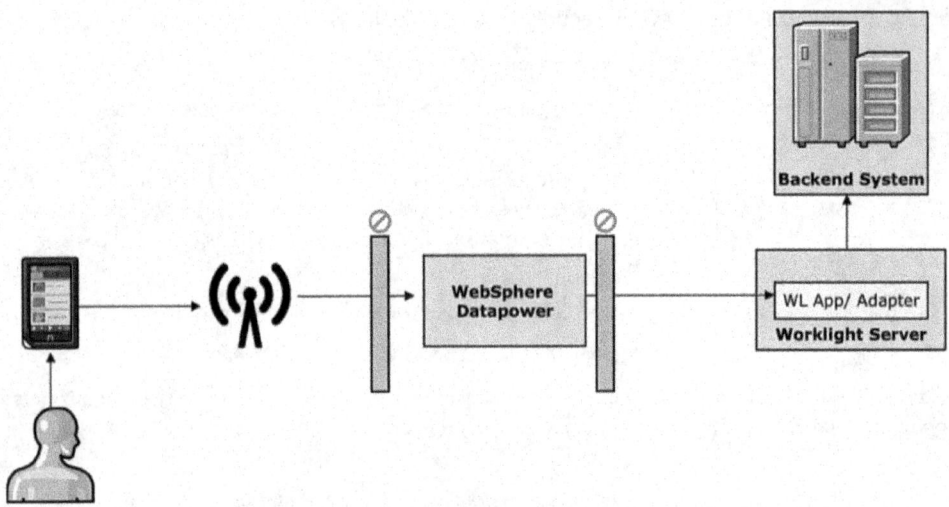

Figure 5.4   IBM DataPower appliances authenticate the mobile user and protect the communication to the mobile application's server-side component

When the mobile app running on the device needs to connect to the server, such as for data or services, the request is routed to the DataPower Appliance, which is in the DMZ. If the request does not have user authentication credentials associated with it, the appliance challenges the mobile app to supply the user credentials (e.g., user ID and password) and then authenticates the request against a user registry.

When the request is authenticated, the appliance injects a token into the HTTP header (identifying the user) into the request, and sends that request downstream to the MEAP server that has been configured to trust that the DataPower Appliance can assert the user's identity. See Figure 5.5 for a typical interaction between the user, the mobile application, and the MEAP server.

For a complete list of the currently available IBM WebSphere DataPower appliances and their functions, view the product details at http://www.ibm.com/software/integration/datapower.

Mobile Devices Security Considerations

In a mobile world, there can be little or no control over the device in terms of who are using it, when and where it is used, and what it is used for. This presents some unique challenges to the mobile application development teams.

In certain scenarios, this is mitigated by deploying precustomized devices that are preloaded with sanctioned applications and stripping away system privileges from the user so that no customization can be done on the device and no additional applications can be installed. This means employees are carrying around a special device for work, which is dedicated to performing certain work-related tasks. For instance, airline pilots are starting to use tablet devices loaded with

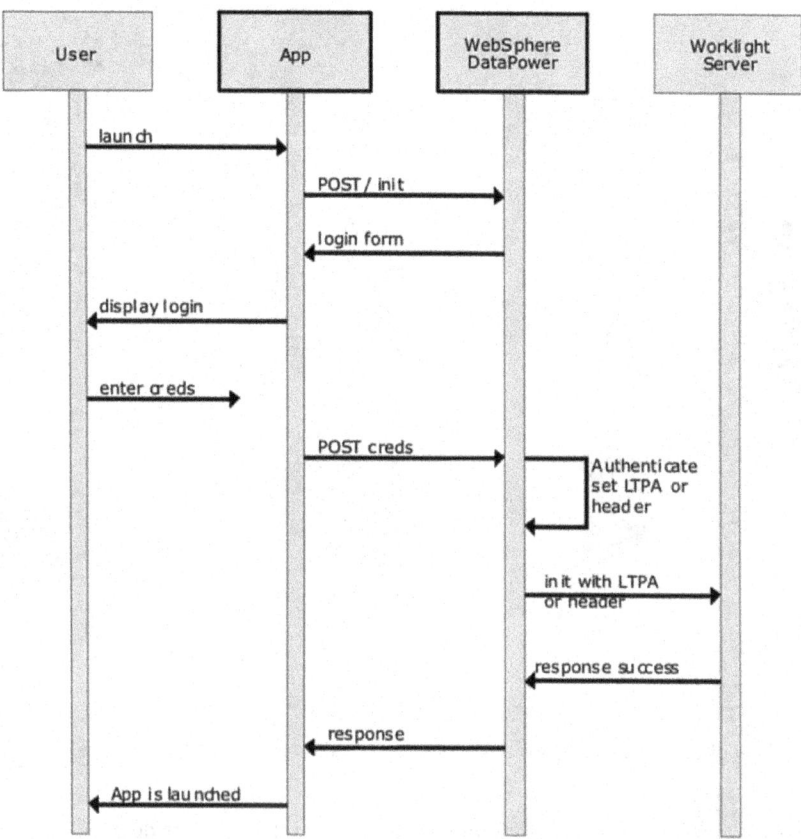

Figure 5.5 A typical interaction between the mobile application and the MEAP server

flight data, manuals, and aviation maps that replace the bulky paper documents. However, this practice can only be limited to certain use, often applicable to only highly specialized lines of work. To most of the enterprise use cases, BYOD offers much better return on investment and is more practical to widely deploy the enterprise mobile applications to the employees.

When an enterprise adopts a BYOD policy, to provide the ultimate convenience to their employees and save the enterprise from having to manage a large fleet of company-owned devices, it is necessary to ensure applications and enterprise data are properly managed and protected (Figure 5.6).

In general, mobile security challenges fall into these categories:

Loss and Theft

The practice of BYOD encourages the mix of private and personal use on the same mobile device, which makes loss and theft the most obvious security concerns because these devices may contain business data of varied levels of sensitivity and can be used to connect to enterprise networks.

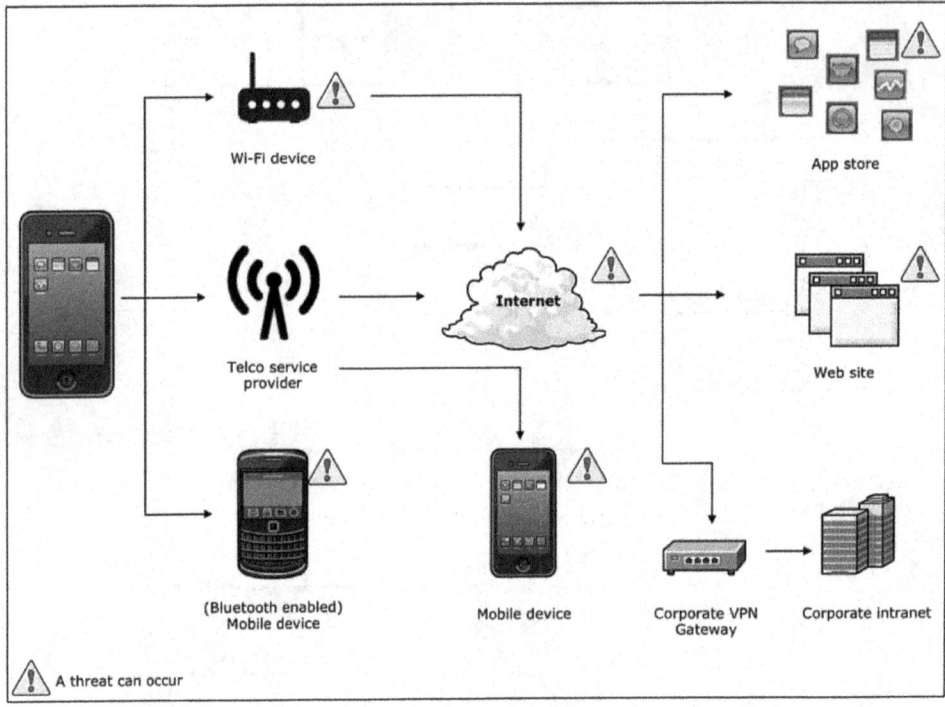

Figure 5.6   Flow of data transmission and potential locations of exploitation

According to a mobile thread study by Juniper Networks, 1 in 20 mobile devices has been stolen or lost in 2010.

While it is impossible to prevent a mobile device from being stolen or lost, a number of security measures can be employed to minimize the impact of a stolen or lost device:

> **Using a complex password**—Personal phones are often not protected with a password if the user does not want to bother with typing the passcode everything he or she wants to use it to check a friend's post on social network or make phone call. While essentially all device manufacturers provide an option to secure the phone with a passcode, they are typically weak. For instance, iPhone's default passcode is only four digits. In order to elevate the passcode strength to meet the enterprise's security standard, a custom security profile is typically installed by an enterprise mobile application and device management software as the first line of defense against loss of corporate data. For example, when installing the mobile client application for MaaS360®, part of IBM's MobileFirst portfolio that helps mobile IT organizations to manage mobile applications and devices, a security profile is installed which requires the phone's passcode to be a much more complex string of alphanumeric characters.
>
> **Locking the device from being able to access the enterprise networks and discontinuing application usage**—Mobile security management software often requires

devices to be registered, which allows IT operations to track the user of the device and the enterprise applications installed on it. This way, the administrator can block a device from using the application backend or accessing the enterprise network. In addition, the mobile application can be developed in such a way that it always checks with the mother ship during startup for legitimacy, using factors such as the mobile application's authenticity, presence of a secret client certificate.

Wiping data remotely—Mobile security management software must support remote data wiping. In the event that the device is lost or stolen, sensitive or confidential enterprise data can be remotely cleaned up from the device as long as a connection can be established with the device.

Malware

Malware includes viruses, worms, Trojans, and spyware. They have existed for almost as long as the Internet itself. Malware targeting mobile applications has been on the rise. Some of the consequences of malware infection include the following:

- Cause the loss of personal or confidential data.
- Incur additional service charges by sending premium SMS messages or initiating phone calls in the background.
- Render the device unusable.

While none of the mobile platforms provide built-in malware protections, the situation on iOS is better because of Apple's vetting process on all AppStore submissions, which can successfully detect malware during the review process and at the same time acts as a deterring factor. On the Android side, on the other hand, malware exploitation has been steadily increasing because distribution of Android applications is open to anybody and there is no requirement to use a digital certificate that is properly signed by a Certificate Authority. An Android marketplace, including Google Play, does not typically require application submissions to be reviewed. As a result, it is much easier to publish applications with embedded malware for the Android platform. If a user is not careful and installs applications from an unknown source, the device can easily be infected by malware. According to the research report from Juniper Networks, malware on Android has increased by 400% from June 2010 to January 2011.[2]

Companies can significantly reduce the malware risk by adopting a similar approach to be used for both mobile devices and the desktop and laptop environment. In addition to advising employees to only download and install trusted applications, and take appropriate actions when suspicious applications are identified, a company should run antimalware software on each employee's device to detect malware in real-time and scan the entire device periodically. This is typically part of the Mobile Device Management strategy, which will be discussed in a later part of the chapter.

---

[2] Juniper Networks, 2012. *2011 Mobile Threats Report.*

Phishing

"Phishing" is an email or an SMS text message written in such a way that tricks a user into accessing a fake website impersonating real commercial web sites, sending a text message or making a phone call to reveal personal information (such as a Social Security number in the United States) or credentials that would allow the hacker access to financial or business accounts. Phishing through mobile browsers is more likely to succeed because the small screen size of mobile devices does not allow for some protection features used on the PC, like web address bars or green warning lights.

The most effective anti-phishing approach helps a user recognize a fraudulent website when it is presented. Some financial institutions have deployed "site authentication" to confirm to users that they are communicating with a genuine website before they enter account credentials from either a web browser or a mobile application. Two-factor authentication is also useful to thwart phishing: First, a user enters a static password, and then a second authentication factor, such as a one-time password or a device fingerprint is dynamically generated to further authenticate the user. So even if a hacker using a phishing technique steals a user's static password, the hacker cannot login to the genuine site without the user's second authentication factor.

Understanding the Worklight Security Integration Framework

IBM Worklight defines an extensible security integration framework to accommodate for different authentication and authorization mechanisms employed by enterprise backend systems.

This process is defined in an *authentication realm* in the Worklight server's configuration, and consists of a *Challenge Handler* on the client side, and an *Authenticator* and a *Login Module* on the server side (Figure 5.7).

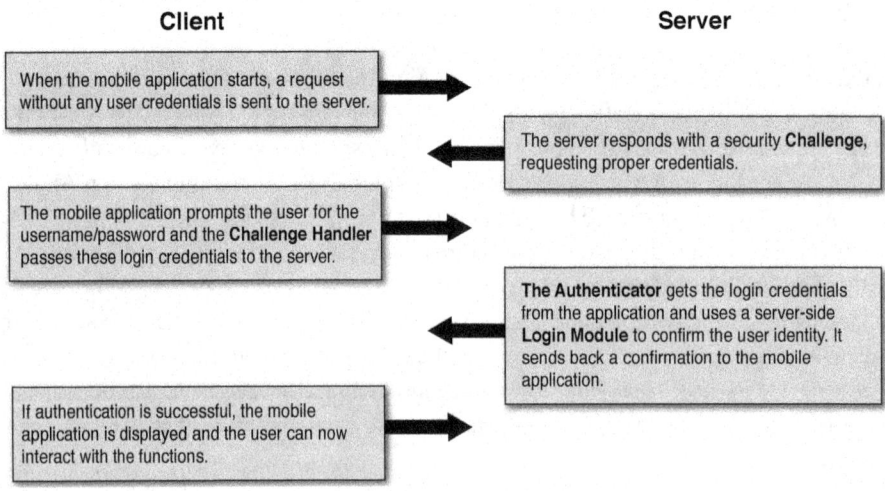

Figure 5.7   IBM Worklight security integration framework

Each time a Worklight application contacts the Worklight server, it will be challenged by the server to present a proper credential to certify its entity. A client-side component called "Challenge handler" then decides whether the current user has already been authenticated. Depending on the exact authentication mechanism used by the application, the existing authentication can be saved in the form of a session cookie or a client certificate. If one already exists, the challenge handler will present that back to the server. Otherwise, the user will be prompted to log in.

The challenge handler, running in the mobile application on the mobile device, will pass the login credentials from the user input to the server. On the server side, the authenticator collects the credentials from the client and passes them along to the login module. There are different login modules to use depending on the authentication mechanism. For instance, to use enterprise LDAP service to authenticate, use a login module for LDAP authentication. The login module will perform the actual authentication by using the credentials collected by the authenticator. Note that the login module must work in coordination with the authenticator in order to ensure the right credentials are collected from the incoming authentication request. Different authentication mechanisms require different credentials. For instance, WebSphere LTPA uses a special token embedded inside a cookie to preserve the authenticated user session. For the mobile application to work with LTPA, an LTPA-aware authenticator is needed to extract the token from the incoming request and pass that along to the login module.

Once the login module validated the user credentials, the success is acknowledged in the HTTP response. An authenticated user session is established to allow the user to access the protected server-side resources through the mobile application. The Worklight mobile authentication process described above is illustrated in Figure 5.8. For more information on the Worklight security framework, and integration with enterprise security infrastructure, visit the IBM Worklight Foundation 6.2 Information Center.

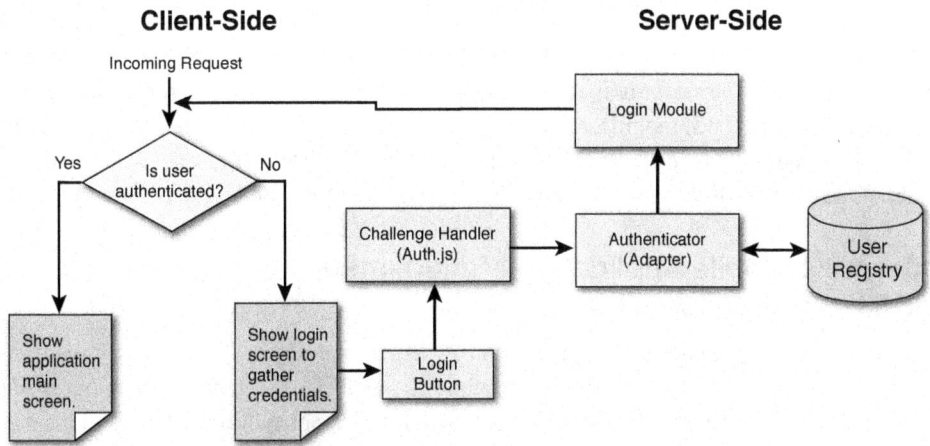

Figure 5.8 End-to-end authentication flow

Secured Data Store and Synchronization

A lot of enterprise mobile applications will be required to work in off-line mode, either due to the working environment without Internet connections, or for the application to support undisrupted usage with unreliable networks. If the application consumes or produces data, working in off-line mode means the data must be available locally when the connection is dropped, and later on when the connection is restored, the local data will be automatically synchronized with the mother ship.

Local data storage typically needs to be encrypted to protect confidential corporate data or sensitive personal data from unauthorized access.

A typical way that data synchronization works is when the mobile application first contacts the server, a copy of the data stored in the server is downloaded and stored on the device. The application can continue to have direct contact with the server for data or start using the data in the local storage. When data is updated in the server, the user is either prompted by the application to refresh or refresh is only done when the user explicitly requests to re-synchronize. How the data is refreshed depends on how important the application has up-to-date data locally. For a pilot flight plan and operation manuals application, it is critical that the data is synchronized before the flight takes off to ensure the airport map and information on the flight route is accurate. On the other hand, an application used by miners down in a coalmine to manage the equipment has relatively stable data that do not change as often.

The tricky part of data synchronization is when the mobile application submits data back into the System of Records. When using local data store, new records may be created and saved in the data store because a direct connection is not available at the moment of the submission. Because the application may have operated on stale data, for instance, picking an equipment part number that became out of stock since the last time the data synchronization was done, synching the data stored locally back to the master database requires careful processing. This is a problem that any distributed database must address. Using a local data store essentially makes it a distributed database architecture. A technique used by many such databases, such as IBM Cloudant®, is for Update and Delete requests to include a _rev field, which is an internal counter issued by the server to mark the revision count of the record. The server uses the revision field to figure out if the submitted record from the client application has been modified since it was retrieved (at which point the internal counter was generated). Some popular enterprise MEAP and cloud-based MBaaS offer a device-local data store, as part of the client SDK that automatically synchronizes with the mobile data backend.

Enterprise Mobile Application Management and Device Management

Enterprise mobile applications are part of the IT assets that need to be carefully managed for security and audit reasons. Managing the "mobile channel" can be achieved by either managing the devices that run the mobile application, or managing the applications and users of the application. Typically, a combination of the two approaches is needed in order to meet the enterprise IT compliance requirements.

Distributing and managing mobile applications in an enterprise context is a crucial part of a comprehensive mobile strategy. Most of traditional application management principles apply to Mobile Application Management, such as distributing the applications to the employee's "end-points," or mobile devices, rolling out updates and mandatory security patches, notifying on the employee devices of important events such as operating system upgrades. These are typically accomplished through installing an application management client on the employees' mobile device that will enroll the user and the device, in order to allow the management server to accurately target the appropriate users and devices for specific system operations.

Special Challenges in Managing Mobile Applications and Devices

In addition to these basic principles common to any endpoint management systems, a Mobile Application Management system must also deal with special challenges that mobile devices present.

Data Loss Prevention

In today's practice, it is no longer possible to issue company-owned devices that have strict security policies enforced. Instead, most enterprises choose a BYOD policy. In such environments, enterprise applications and applications installed for personal use are sitting side-by-side on the same device. The same application, such as the Internet browser, may also be used to conduct both business operations and to use for personal leisure. Data leaks can happen when careless employees copy confidential information from a business application and paste into their personal emails or chat messages, or take a screenshot of an architecture diagram of the company's next generation product and share with a social network application. An enterprise Mobile Application Management strategy needs to support Data Loss Prevention to avoid these situations.

Remote Wipe

Companies should also be able to remove applications and data from a registered device when necessary, such as when the device is lost or stolen, or when the employee has left the company or changed job responsibility. This procedure is often referred to as "remote wipe." A remote wipe can either be requested by an employee, for instance, if the device gets lost, or initiated by the mobile operation administrator on noncompliant devices.

Geo-fencing

Furthermore, application security policies can also be tied to mobile devices' geo-locations. This is a practice often referred to as "Geo-fencing." A hospital may secure the patient care application used by doctors and nurses so that it is only operational inside the perimeters of the hospital building. This will protect the sensitive patient data from leaking and prevent the application from being misused and putting the patients in harm's way.

Mobile Device Management

In addition to managing the application usage, the devices themselves often need to be managed. This includes requiring a more secure passcode to unlock the device to ensure certain business applications do not fall into the wrong hands. Companies may also mandate security patches or upgrades released by the device manufacturer or the operating system vendor, which may otherwise be ignored or delayed by the employees.

Usage Reporting and Analytics

A mobile application and device management system is also the ideal place to track usage patterns and trends. Most of today's mobile management software generates automatic reports on the distribution of the mobile operation systems across the user population, device capabilities and ownership (BYOD or company-owned).

Example Product: IBM MaaS360

In the IBM's MobileFirst portfolio, MaaS360 is the enterprise mobility management platform, providing both Mobile Application Management and Mobile Device Management capabilities. It can manage mobile apps that are internal deployed or published to the public app store, all from a single administrative portal. For Mobile Device Management, MaaS360 can streamline the provisioning of company-owned and employee-owned devices over-the-air. It takes the users through the process of device enrollment and security policy configuration with a smooth user experience. It allows the enterprise mobility administrators to manage security policies and carry out device actions such as locate, lock, and wipe.

For Mobile Application Management, MaaS360 simplifies the lifecycle management of mobile apps for distribution and updating. The app can be private, public, and purchased. It offers an easy-to-use enterprise app catalog with full security and operational lifecycle management across mobile device platforms.

To protect against data leaks, MaaS360 offers the following apps:

- A secure mail app with email, calendar, and contacts for iOS, Android, and Windows Phone devices

- A secure browser for secure access to intranet sites and web apps, and automated compliance of content policies for iOS, Android, and Windows Phone devices

- MaaS360 Mobile Application Security as a mobile application container with full operational and security management to protect against data leaks for iOS and Android devices

Enabling a mobile application for secure management can be achieved either at application development time, by using special libraries or SDKs that enhances the security of normal system operations, or by rebuilding the application during deployment time and replacing certain system libraries with special libraries that provide enhanced security. The latter approach is often referred to as "application wrapping." Mobile application management products that support application wrapping typically allows an administrator to upload an already built mobile application binary (.ipa for iOS or .apk for Android) to an administration port, and select the list of security policies to enforce on the application, then kick off a special build that replaces the relevant libraries within the mobile application binary.

The application wrapping technique allows the enterprise to centralize the control of mobile application behaviors according to the corporate security policies, without having to ask each application development team to follow the same practices to develop with the necessary security libraries. On the other hand, typically using the SDK, development teams have more flexibility on the behaviors of the application in order to build the best user experiences. All of the security requirements can be met by using either technique. For example, file I/O can be enhanced to require stronger encryption than provided by the mobile operating system by default, such as the Federal Information Processing Standard (FIPS) 140-2. Network communications can be intercepted and redirected through a corporate VPN tunnel to protect the data in transit. Cut-and-paste from certain applications that contain sensitive information, such as a financial analytical services app, can be disabled by applying encryption to the clipboard, resulting in garbled content being pasted into an unmanaged application that does not have the right security key to decrypt the clipboard content.

Architectural Choices for Secured Enterprise Connectivity

From the architecture point of view, when connecting to enterprise backend systems from a mobile application, an adapter layer is needed. Adapters sit between the mobile applications and the APIs of the backend systems. These are either custom-developed server-side components for a mobile solution, or commercial off-the-shelf integration solutions.

Let us first evaluate using adapters that are custom-made for one or more mobile applications.

Why are adapters a good idea for mobile applications consuming enterprise backend system services? First of all, adapter is not a new concept. Many existing software solutions that involve multiple systems often employee adapters for integration purposes and keeping the impact on the whole solution to minimum one or more systems are changed. In other words, adapters promote decoupled architectures. In Java Enterprise Edition (Java EE), Enterprise Java-Beans session beans are one kind of adapters.

There are more reasons to use adapters in an enterprise mobile application. In today's complex IT systems, it is rarely the case that the backend APIs needed by the mobile application are directly consumable. Maybe the APIs were designed to support a wide-range of usage patterns but the mobile app only needs to use a small subset, which as a result requires the input parameters and return values to be simplified to improve efficiency. Adapters can optimize the invocations to the backend APIs and simplify the calls from inside the mobile applications, which results in faster network communication and lower bandwidth consumption.

Or, maybe the protocols are geared toward machine-to-machine processing and emphasize grammatical accuracy, like XML, but are overkill for UI-to-machine communications. For example, most UI applications provide client-side validations to ensure user input conform to the data model on the server side. However, if XML is used to transfer data, when the server processes the data using an XML parsing library, an XML schema is often required to do the transformation from the markup content to internal data models. In this process, the user input is validated again, using the XML schema. In most cases, the validation on the server side is not needed but is unavoidable with XML.

In almost all mobile applications, JSON is the preferred data transfer format because of its compactness and fast parsing. Adapters can do the protocol transformation between JSON, which is needed by the client side, and other protocols supported by the backend APIs such as XML (including SOAP).

In IBM Worklight, which is the mobile application platform for IBM's MobileFirst portfolio, adapter is a standard architecture component for developing mobile applications. Worklight Adapters connect to enterprise backend systems and deliver data to and from mobile applications. Worklight Adapters also perform server-side logic that is specific to the mobile applications they serve.

Worklight Adapters are deployed to the Worklight server and can be access from any mobile applications deployed to the same Worklight server (Figure 5.9).

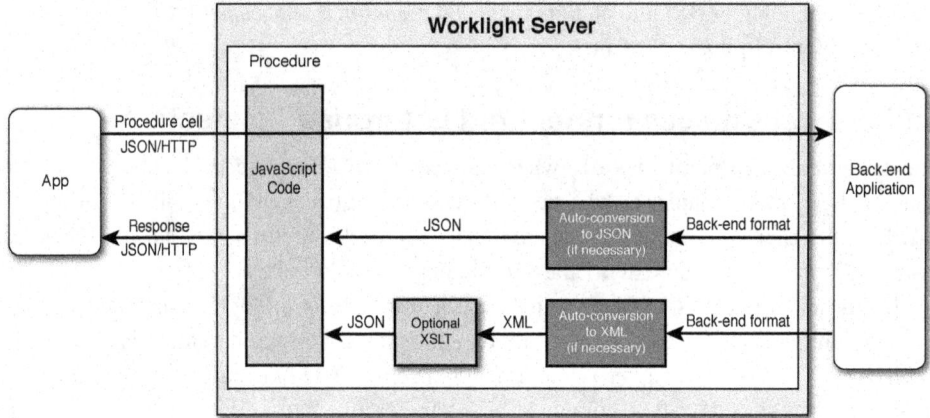

Figure 5.9   IBM Worklight Adapter framework

Worklight Adapters are developed in JavaScript. To assist in transforming XML data from the backend APIs to JSON, Worklight Adapters support XSL that the developer can supply to make it easier to parse and process hierarchical data.

When adapters communicate with the backend systems, they can assume the security identity of a predetermined user account, or the user account of the incoming request from the mobile application. The former is called "run as server." The latter is called "run as user." Which connection policy to use depends on what options the target backend system supports and the access privilege required by the mobile application. Another factor to consider is billing options. A backend system may charge per user account. You may need to use "run as user" if the users of the applications are expected to shoulder the cost of utilizing the backend system's services. Or if this cost can be covered by a functional account that serves all users, then select "run as server" and configure the adapter with that functional account.

What about using commercial solutions that are ready to use with minimal configurations to support a variety of enterprise backend systems?

WebSphere Cast Iron Live is the cloud integration-as-a- service solution. Using a "develop once, deploy anywhere" approach, WebSphere Cast Iron Live is ideal for customers with a majority of their applications based in the cloud and with no infrastructure on premise. The offering follows the same model as Software-as-a-Service (SaaS) or On Demand services. SaaS approaches run a company's business applications through a network on a remote host and look and operate exactly as if they were running on the company's own systems. WebSphere Cast Iron Live runs under the same model, meaning companies who integrate using this product can integrate their SaaS and web-based applications in real time.

WebSphere DataPower Cast Iron Appliance XH40 is a stand-alone, self-contained hardware offering. It is the preferable option for customers with a majority of applications based on-premise who need a standards-based solution and who find software-based integration solutions to be too complex. It comes with all of the required programming on-board for a particular integration project. The device is called an "appliance" because it has the same self-contained/dedicated function characteristic as most appliances, like a network router. They look like any other rack-mounted box, but are dedicated to one important task: integrating multiple on-premise or SaaS applications.

IBM WebSphere Cast Iron Hypervisor Edition is a virtual appliance for service integration. It provides connectivity to a variety of cloud-based and on-premises applications, databases, web services, messaging systems, and other endpoints. In addition to predefined endpoint connectors, IBM WebSphere Cast Iron Hypervisor Edition includes templates that enable access to other endpoints such as Google Analytics, Microsoft Azure Servicebus, or Oracle CRM on Demand.

IBM WebSphere Cast Iron Hypervisor Edition uses a configuration approach in which integration projects are configured instead of programmed. With a configuration approach, mobile applications can be connected to backend systems without any help from a developer. Instead, you build the needed integration flows using a graphical development environment in WebSphere Cast Iron Studio.

Each integration project consists of one or more orchestrations, each of which describes a specific set of activities that define the flow of data, including access to backend systems and data transformations. Mobile applications can use an IBM Worklight Cast Iron adapter to perform the activities defined in an orchestration.

Transforming data from its original format to another format can be done using a simple drag-and-drop method in the WebSphere Cast Iron Studio graphical mapping editor, which is shown in Figure 5.10. There is no need to understand all supported data formats because the data transformations are done visually using the mapping editor.

Business logic rules also can be set up using the graphical mapping editor. Because it is a WYSIWYG (what you see is what you get) interface, business users can participate in an integration project without any specific IT skills.

For administration, WebSphere Cast Iron Hypervisor Edition provides a web-based console to monitor the data flow, handle exceptions in the data flow, and provide proactive alerts regarding data and connectivity errors. The console also provides easy access to key performance indicators.

IBM WebSphere Cast Iron provides a secure environment with capabilities to easily and rapidly assemble and publish useful business services. Reusing these existing business services (and their existing infrastructure) can reduce a new application's time-to-market. A company can also offer these business services to other companies to open new market opportunities.

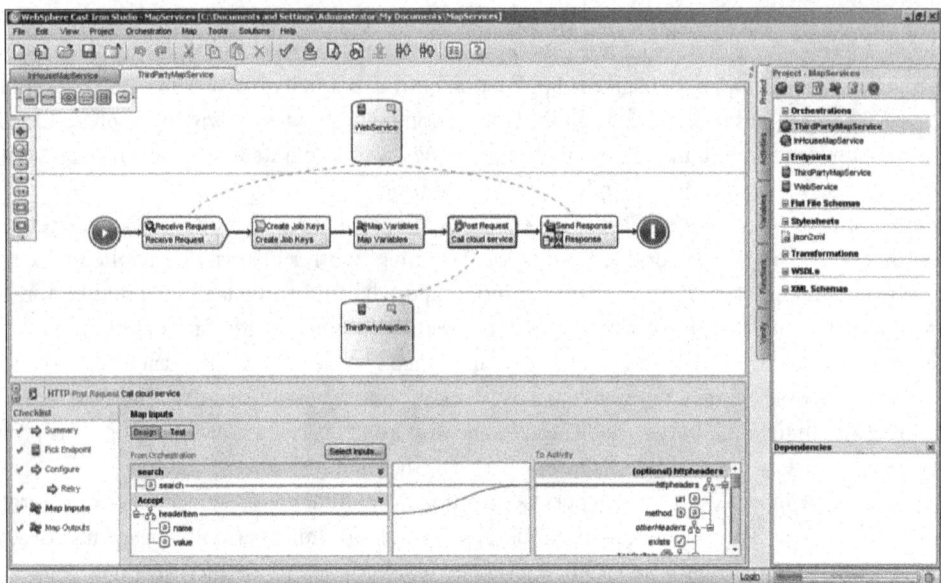

Figure 5.10   IBM WebSphere Cast Iron Studio graphical mapping editor

More information about IBM WebSphere Cast Iron can be found at http://www.ibm.com/
software/integration/cast-iron-cloud-integration/features.

Summary

Mobile applications for both customers and employees are increasingly becoming a center of
gravity for modern enterprises' innovation and digital transformation. In the journey to mobile
digital excellence, enterprise IT and lines of business departments alike are both facing unprec-
edented opportunities to make fundamental changes to the company's business model. At the
same time, the digital transformation led by mobile is a huge undertaking. New customer scenarios must be studied and developed. Existing backend systems must be adapted to meet the new
workload requirements. Continuous user experience across multiple channels is the key to success. Security on all levels from the device management to data storage to message transmission
to backend access control must be carefully planned and implemented. Good news is that in the
competitive MEAP market, vendors are offering increasingly mature and innovative solutions to
help make the journey a smooth one.

When innovation user experience design on the device side is paired with highly available, speedy, and secure backend services, killer mobile applications will emerge and capture the
precious user attentions. The best mobile solutions do not stand alone, but instead work seam-
lessly with the other UI channels and provide a continuous user experience, no matter where the
users are.

CHAPTER 6

A Comprehensive Approach to Testing of Mobile Applications

This chapter will focus on enforcing quality in your mobile applications. It will explore the various aspects of mobile application testing and the environments that would be necessary to facilitate the testing of these applications. It will go into detail on the what, why, and how of testing a mobile application. Assessing and enforcing quality in a mobile application belongs throughout the development lifecycle and below you will find a description of what type of testing is fruitful at which stage of the mobile application development.

Why Is Quality Essential?

As we have discussed so far, design and implementation choices of building mobile applications are very critical for the overall success. However, having a maniacal focus on quality of the mobile application is equally, if not more, important and critical for the application being accepted by our customers. Adopting test driven development with agile methodology and automated DevOps (as discussed later in this book), empowers the application owner to focus on the unique selling proposition of the mobile application. Due to the imminent possibility of a quick trial and an even quicker rejection of a low-quality mobile application, the development team may find itself constantly addressing concerns over its functionality, performance, and feedback from various adopters in the Application Store.

When one plans for testing effort of the mobile application, the various aspects of the mobile application should be kept in mind. There is the

- mobile frontend that might be available on various operating systems and form-factors
- middle tier on a server infrastructure, and
- the mobile backend

A comprehensive approach to testing all aspects of the application environment is critical to ensure reliable end to end quality.

Most often, the entire infrastructure for the application environment may not be available at all given times, hence forcing a waterfall nature of development. There are some creative ways of getting around this to speed up development and testing of the application, before putting it out for restricted or general use by leveraging virtualization techniques.

A challenge is to make sure that the application works equally well on all supported platforms and form-factors. While it might be impractical to test on each and every configuration and mobile operating system out there, we shall discuss ways where automation will help reduce the overall cost of making sure the application has improved in quality. There are varieties of testing that a team may adopt to ensure overall quality.

This chapter will provide a perspective on testing of mobile applications while demonstrating how the overall quality could be ascertained using IBM Rational toolset.

When Should Quality Be in Focus?

The appropriate time and occasion to focus on quality is right from the beginning. If you are developing a modern age mobile application, chances are, and it is imperative, that you have adopted an agile method of application development. We have focused on design and implementation in previous chapters, while this chapter will focus on application quality as its being built and tested all the way up to maintaining the quality in subsequent releases of the application. We will describe methods of both manual and automated testing.

What Is the Cost of Quality?

The cost of quality has many aspects and it compounds proportionally to the lack of focus on it. The simplest form is the amount of time (converted to money) spent on the efforts to prevent defects much before the app is released. This is the easy part. Where it compounds is that the cost of not controlling quality during development adds to the cost of formal testing. The cost of not controlling quality during formal testing eventually could lead to customers rejecting the software due to lack of quality. It is a known fact that the later a defect is found in software, the more expensive it is to fix and deliver. It is also a known fact that often times fixing one defect uncovers other hidden defects which are possibly more expensive to fix.

In the case of mobile applications, it leads to acceptance or rejection of the application, and a huge negative impact through bad ratings in the app store. A mobile app with security vulnerability could potentially expose the app to a much higher cost and even more serious consequences.

With accelerated release timelines and the potentially high cost of quality for a mobile application, the challenge is to create an application that is both accepted and adopted. In this chapter, we will discuss some of the quality related steps that an app developer should take towards this goal.

Automated versus Manual Testing

Some mobile app testing is performed in an automated, unattended manner, while other testing needs to be done in an interactive, manual style. The best comprehensive quality assessment of your mobile app will employ a balanced combination of both automated and interactive testing.

Automated mobile app testing is critical to accelerate delivery of your application and maintain the velocity of your mobile app development lifecycle. There are a wide variety of automated testing techniques that can be applied to mobile apps. Each technique has certain strengths and it is important to strike a balance across the forms of automated mobile app testing.

- Random generated mobile tests (aka "monkey" testing)
- Key word based mobile app test scripts
- Programmatic user interface (UI) testing applications (UIAutomator/UIAutomation)
- Behavior Driven Development (BDD) testing
- Image recognition based automation
- Instrumented application object/event based automation

We will discuss the pros and cons of some of these automated techniques in subsequent sections of this chapter, but next consider how manual interactive mobile app testing fits into your quality regimen.

Automated testing of your mobile application is not sufficient to ensure the best quality app. There are aspects of the quality of your app that cannot be determined using automated testing techniques alone. The "look and feel" of the application, its usability, the logical flow of the user journey through the application function—these are some of the aspects of a mobile app that are more subjective in nature, and therefore, are a better fit for human-executed interactive manual testing and assessment.

A best practice of mobile app testing strategy strikes a balance between automated tests and interactive human based testing (see Figure 6.1). The ideal quality cycle begins by running a battery of automated tests against the output of the continuous integration build process for your mobile app. Once this initial battery of automated tests has verified that the latest build meets minimum quality criteria, the build can be distributed to a group of testers/internal evaluators who will perform the interactive testing.

After both automated and interactive tests have passed for the mobile app, it is a candidate to be released to production use and distributed to real end users (presumably via public app store or else some private enterprise app store). Even after the mobile app has been released into production, you can continue to obtain quality assessment data about the application.

Preproduction versus Postrelease

The main focus for quality assurance of mobile apps is on the preproduction phases of the development lifecycle. However, quality assessment about the app should not end when the app gets released into production. There is still very important data about the behavior of the mobile app "in the wild" that can be obtained and used to help the developers continuously improve the app.

It is not practical to expect that every conceivable defect in a mobile app can be caught and fixed before it is released into production. Even after deployed to the app store and installed on end users' devices, the best mobile apps continue to capture context information for every crash that occurs and deliver that information back to the app development team. It also helps

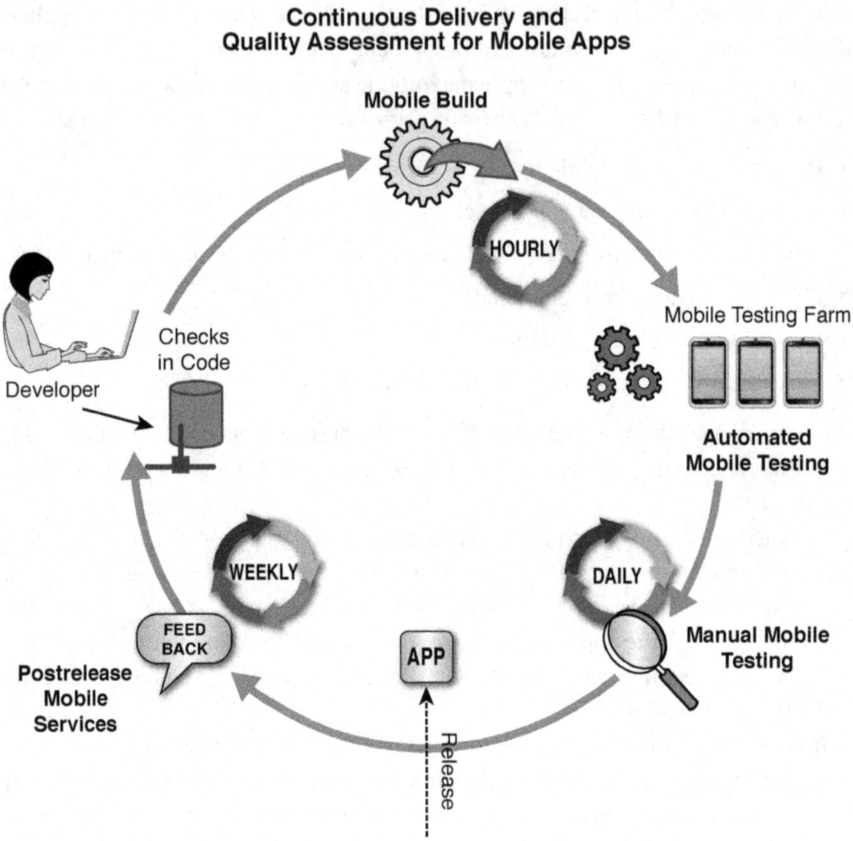

**Continuous Delivery and
Quality Assessment for Mobile Apps**

Mobile Build

HOURLY

Mobile Testing Farm

Checks
in Code

Developer

**Automated
Mobile Testing**

WEEKLY

DAILY

**FEED
BACK**

APP

**Manual Mobile
Testing**

**Postrelease
Mobile
Services**

Release

Figure 6.1   Ideal mobile app quality cycle

discover the most often used portions of the application and the user experience related to them. This is one method of discovering potential enhancements in the application that would be most effective.

Many times, a crash only occurs under very specific conditions that are difficult or impossible to recreate in the test lab. The problem could arise from the use of an unusual or old mobile device model, or from special network conditions, or the combination of certain other mobile apps running concurrently with your mobile app. These real world circumstances are impossible to anticipate and impractical to cover in all permutations in preproduction test time frames. So your development organization has to depend on receiving good technical contextual data when they occur in the field, so that the root cause of the crash can be quickly determined.

Besides outright crashes, it is valuable to solicit feedback from your end users about how they perceive the mobile app. Most popular mobile apps include some kind of "in-app feedback" mechanism so that users who would not take the time and effort to write a review in the app store can at least send the development team a short message about how they view the app. And it is especially important to capture the context of such feedback at the time it is submitted, so that the development team can know if special conditions are contributing to the impression conveyed by that user in their feedback.

Automated Mobile App Testing Considerations

There are several aspects of automated mobile app testing that bear special attention, including the devices used for testing, isolation of the code running on the mobile device, and the specific technique employed to create and execute mobile test automation.

Test Devices

A crucial consideration for automated mobile testing, regardless of the type of testing to be done, is the type of mobile device (or devices) on which to execute the automated tests. The automated tests could be executed on a small number of tethered real physical mobile devices. Or they could be executed on emulator programs running on the developer's workstation. Several vendors offer a remote "mobile device cloud" as a potential target for test execution. Device emulator programs could even be scaled up in a virtual device cloud to provide elastic compute capabilities as testing load varies between peak activity times.

Emulators and Simulators

Emulators come with all of the native mobile operating system development kits, and simulators are available from several sources (including IBM). Emulators attempt to replicate the actual mobile operating system running on top of some other hardware, such as a PC workstation. Simulators do not attempt to replicate the mobile OS, but instead provide a light-weight simulation of the UI.

Using emulators and simulators for some amount of testing the mobile app can be cost effective, especially, in early stages of code development. The typical code/deploy/debug cycle especially for simulators is much more rapid than for physical devices (typically) and use of these tools eliminates the need for a developer to have the real physical device in hand.

However, there are subtle differences in behavior between device emulators and real physical devices (even for the very best emulator programs), and simulators do not allow execution of some parts of the application logic flow (only the UI look and flow). So, while emulators and simulators can be used to cut costs and speed development, they are generally not acceptable as the only form of test execution for mobile apps.

IBM offers mobile simulators as part of the development tools for the IBM mobile enterprise solution. Emulators are always supplied directly by the supplier of the mobile operating systems (Apple, Google, Microsoft, RIM, etc.).

Device Clouds

Broad-scale testing on real mobile devices is crucial for any app that will be released into the consumer market or within organizations that have opened up for a bring-your-own-device strategy.

There are several approaches for on-device testing including the category called device clouds. What about the problem of the sheer number of different physical mobile devices on the market? There are literally thousands of different device types running different release levels of mobile operating systems, and connected to different network carrier providers and wireless networks. The combinatorial complexity of the universe of possible permutations is almost beyond comprehension. The cost of owning, setting up, and managing all of those different combinations is completely prohibitive even for very well funded projects.

A technique that can address this problem is to employ a "device cloud" testing solution. Device cloud is a term used to describe a very large array of real physical devices that have been made remotely available for access across the Internet much in the same way that general compute resources are made available in a generic software "test cloud" solution.

The test organization arranges to reserve some mobile devices for test for a certain amount of time, and deploys the mobile app code to the devices where automated tests are run using whatever "on device" automated test solution is the choice of the test organization. Once the current testing cycle is completed, the reserved devices are relinquished back to the "device cloud" where they are available to be used for other mobile app testing, potentially by a completely different project.

This technique does not eliminate the need for manual interactive testing by humans, and in fact works best in conjunction with some form of execution of those other techniques. What this approach is good at is reducing the cost of ownership for the huge variety of device types that exist and can be expected to be employed by the users of the mobile app once it gets into production.

A test organization can invest in purchasing just a few key mobile devices and "rent" the rest of the combinations from the "device cloud." The same automated techniques for mobile testing used on the stand-alone physical devices can also be used for the devices in the "cloud," so results of the automated testing are consistent. While it handles the array of devices to test on, this approach does not provide for testing connectivity states other than turning off the connection. One of the most overlooked testing scenarios is what happens if the device disconnects during various key uses of the app. Some services do offer the ability to select which carrier the device is running on, but ideally tests should be run in as close to real use scenarios as possible.

There are issues related to this kind of testing, whether the resources in the cloud are general compute resources or mobile device resources. Issues of security of the app under test, public or private device cloud, and balancing the cost of the cloud with the potential cost of the defects eliminated are all issues that any cloud testing solution has to address.

IBM does not offer its own device cloud solution. Instead, we offer integration between our overall mobile testing management solution and a variety of business partners who have device clouds.

Crowd-Sourced Testing

Many organizations find it challenging to get feedback from their own internal users for prereleases of their mobile applications. As a result a set of companies have brought to market solutions that allow for distribution of an application to live testers out in the field.

What is interesting about using these testing services is that you will get real user behavioral feedback. You can also test in region where you will deploy. While it does not necessarily work very well, users can even run through use scripts to ensure anticipated use is specifically tested.

Crowd-source testing introduces the same security risks mentioned for the device clouds so if the application is accessing sensitive production systems or data that should not be public, this approach should not be considered.

Using Service Virtualization to Isolate Mobile Code

Because mobile applications are multi-tier architectures, the process of setting up the infrastructure to support test execution of the code on the mobile device can be time-consuming and costly. All of the middleware servers and services need to be up and available, and typically it is not acceptable to use real production servers for testing purposes.

In addition, many test case failures can occur not because of defects in the code under test, but instead because of problems in the connected components of the application running in other tiers. In other words, if the middle-tier app server has a problem, the mobile device accessing it will fail its test case.

Cost and deployment delays can be eliminated through use of solutions that effectively replicate connecting components of the multi-tier system so that the testing can concentrate narrowly on the code executing on one specific tier of the app. By leveraging solutions such as the service virtualization capability in IBM Rational Test Workbench, test teams can avoid the need to set up complex middleware environments in support of test execution for code running on the mobile devices.

Service virtualization can emulate the middle tier and backend services and protocols so that the test execution can concentrate on the client tier of the mobile app that is running on the device itself. Conversely, the middle-tier components of the mobile app need to be validated also, and service virtualization can emulate the protocols delivered by the mobile device clients so that tests can be focused solely on the middle-tier functions and services, without need of coordinating physical mobile devices.

Mobile Test Automation Techniques

Staying ahead with mobile apps means frequent iterations with new features. As companies create more updates for their apps, testing quickly gets out of control when trying to do everything manually.

Ultimately there is a requirement to augment and accelerate manual functional verification with some form of automated testing of the code that is executing on the mobile device. This area of the software testing market is new and evolving. There are a variety of approaches that have been created by different vendors. Some of these approaches for automated function test are better suited to typical mobile business applications than others.

Mobile App Programmatic Instrumentation

A typical approach is to place some kind of additional code on the device where the automated testing is to occur. This code acts as a local "on device" agent that drives automated user input into the application and monitors the behavior of the application resulting from this input.

The instructions for telling the agent what to input into the app are typically formatted as either a script or an actual computer program (for instance, written in Java). Creation of these automated test instructions usually requires some proficiency in programming, and many test organizations are short on such skills. Furthermore, creation of these automated test programs is a development effort in its own right, which can delay the delivery of the mobile app into production.

IBM's point of view is that the creation of automated mobile function test scripts should be possible for testers who have no programming skills at all. The tester should be able to put the application into "record mode" and interact with the app normally while the testing solution (i.e., the test agent running on the device) captures the user input from the tester and converts it into a high-level automation script. Once the tests have been "recorded" into a language that is close to natural written instructions, these test scripts can be further edited, organized, managed, and replayed whenever necessary.

Furthermore, since the language employed for the captured test scripts is nearly natural human language, the tester can easily read and modify the script to add elaboration and additional verification points to the instructions. If the language is suitably abstracted from the details of the underlying mobile operating system, these scripts can be executed against real physical devices other than the type used to capture the script in the first place.

IBM has produced just such an automated mobile app test solution and delivers it in both the mobile app development environment (IBM MobileFirst Platform Studio) as well as a software testers' solution (IBM Rational Test Workbench). This technique for automating the mobile function tests is quite complementary to the other techniques described in this document, and can be very effectively used in combination with these other techniques.

Random Generated Mobile Tests

Random generated (sometimes called "monkey") tests have the advantage of not requiring any scripting or coding of automation instructions. Instead of executing a precreated automation script or program, this type of testing introspects the mobile app and generates random pathways of interaction with the app. This random input to the app is executed until a fatal error in the app occurs (for instance, a crash or freeze of the app).

The execution of randomly generated automated tests has proven to be quite valuable for quickly uncovering serious defects in mobile apps that would not have been uncovered through typical scripted tests. Any battery of automated tests for a mobile app should include some amount of this kind of testing. It is not uncommon for this kind of testing to uncover defects in the app within just the first few seconds of execution. As the defects in the mobile app get surfaced and fixed, it may take longer and longer to identify problems in the app using the random test method. But that should be considered a good outcome from use of this technique over sufficient time.

Image Recognition Automated Mobile Tests

Another form of automated testing for mobile apps employs the display images from the mobile device and pixel locations on the device screen. The device display image can usually be captured and automatically compared to a known good image for verification. Automated input to the app under test is defined as a set of tap events targeted at a pixel location on the device display rather than at an internal programmatic application object.

The advantage of this approach for test automation is that it is completely agnostic to the mobile operating system and to the technology used for internal implementation of the mobile app. It is more similar to how a real human interacts with and perceives the mobile app. And highly skilled programmers are not required to produce the automation scripts.

The downside of this approach to automation is that the scripts are highly susceptible to display changes in the mobile app. If the location of a particular widget gets changed by the new build of the app, then the script that depends on the pixel location of that widget will be broken. Some vendors employ advanced algorithms that reduce this "brittleness" in the test scripts. But this technique is so sensitive to app display changes, it is best used later in the development process when the amount of anticipated visual changes in the app are minimized.

Making Manual Testing More Effective

Manual testing is the most common approach for mobile testing in use in the industry today. It is an essential element of any quality plan for mobile apps because it is the only technique that currently provides results for the consume-ability of the app. Intuitiveness and consume-ability are crucial aspects of successful mobile apps and, so far, we do not have a mechanism for automating the testing of this aspect of the code.

But manual testing is also the most time-consuming, error-prone, and costly technique for mobile testing. Manual testing can be combined with other techniques such as the aforementioned crowd-sourcing and "device-cloud" techniques so that the costs and time required can be somewhat mitigated. Solutions that organize the manual test cases, guide the tester through execution, and store the test results can substantially reduce the costs associated with manual testing.

IBM offers a hosted, Software-as-a-Service capability designed to make interactive manual testing significantly more efficient and effective. IBM Mobile Quality Assurance (MQA) services begin the process of making your interactive testing efficient with an "over-the-air" app distribution capability that makes it easy for app developers to deliver updates and new mobile app builds to a targeted set of testers directly on their mobile devices.

When a new build of the app is ready (i.e., passes the initial battery of automated tests) the developer can upload the app binary (.apk or .ipa file) to the IBM MQA service and identify the people who should be notified about the availability of this new build. These app evaluators/testers receive an email notifying them about the new build. And when the tester clicks on a link in the notification email, the new app build is automatically downloaded to their mobile device and installed, ready for immediate testing.

The mobile app tester can be confident that they have the correct build of the app to be tested. As they perform interactive manual testing of the app, when they encounter a defect of any kind, they can use IBM MQA "in-app" bug reporting capability to submit a defect right from inside the app being tested on their mobile device.

The tester simply shakes their mobile device and the app being tested will go into "bug reporting mode." This mode suspends normal behavior of the app and allows the user to capture one or more screen images from the mobile app. The screen image can be augmented with annotations made with the tester's fingers (lines, circles, arrows, anything that you can draw with your fingers).

After the screen image is captured, the tester is presented with a text box to be used to describe the defect in words. Once the description of the problem is entered, the tester taps on the "Report" button and the defect information is sent over the network to the IBM MQA service. Along with the explicit information (screen images and text description) form the tester, rich technical details of the context of the mobile app and the device on which it was running are captured and sent as well.

Some of the rich context for each defect captured includes:

- Mobile device type
- Mobile operating system and release level
- Network in use, including carrier or wireless details
- Memory available and in use on the device
- Logging output up to the point when the defect is reported
- Battery level

This detailed technical information is invaluable for helping the mobile app developers troubleshoot the defect and understand the root cause of the problem.

If you use IBM Bluemix DevOps services for defect and work item tracking and management, you can configure IBM MQA services to automatically open work items for each crash or bug report that comes into the service.

Crash Data Capture and Analysis

In addition to the in-app bug reporting capability of IBM MQA services, every application crash is captured by the service logic. Each time the application crashes, whether during preproduction testing or after the app is released into production, the entire context of the application and the device on which it was running is captured at the moment of the crash. This critical "must gather" data is sent over the network to the IBM MQA services where it is analyzed and made available to the development team.

The crash data capture capabilities of IBM MQA services can be leveraged during the initial automated battery of tests on the app, during the manual interactive phase of testing, and even after the app has been released to the app store.

Additional analytics within the service allow crashes that occur at the same spot in the mobile app to be recognized and aggregated so that you can see how many times a crash occurs at the same location in the app logic. This crash occurrence count is important information that helps your development team prioritize which crashes to fix first. A crash that occurs 1000 times is more important to fix before one that occurs only once or twice.

Performance Testing

Performance testing can be viewed from two distinct directions, when it comes to mobile apps. One form of performance testing is to scale up the number of mobile client instances concurrently running and drive large loads of requests against the middleware and server components of the app. This load and stress testing is important to ensure that the services supporting the mobile app will be able to absorb the traffic that mobile apps are capable of delivering.

The other dimension of mobile performance testing involves measuring how efficiently the code running on the mobile device makes use of the device resources. There are many ways in which a mobile app can inadvertently consume too much of the device's resources (memory, CPU, network, battery) and become unwelcome on the end user's mobile device. A mobile app that is a "resource hog" is likely to be perceived to have poor quality and to be deleted from the user's phone.

Load and Stress Performance Testing

Performance testing for load and stress is one of the more difficult testing scenarios. Typically this involves setting up a test harness that will pour sample transactions to your backend at whatever rate you set. For internal enterprise apps, it is more critical to test peaks and maximums, with consumer apps you need those tests plus spikes and lulls to replicate the natural use states that you would see.

Load and stress testing products such as IBM Rational Performance Tester (part of the IBM Rational Test Workbench) include a recording proxy that allows you to capture the traffic patterns between your mobile app and the services it calls across the network. The recorded interactions are stored as a script that can be run in hundreds or thousands of virtual clients in order to apply load to the mobile app backend services.

At the same time that the mobile app backend services are operating under synthetic load generated by the stress testing tool, it is a good practice to execute functional tests against the mobile app itself to see if behavior of the client side changes when the backend services are under heavy load.

IBM's point of view on performance is that direct integration architectures are brittle and do not enable graceful handling of issues related to volume and stress. Leveraging mobile middleware creates a separate tier that can shield the backends from catastrophic conditions.

Mobile Client Resource Metrics

Client side performance is affected by many different aspects from the level of graphics used in an app to poor coding and bad practices. By measuring and monitoring some of the critical areas such as battery use, device resource use (such as getting GPS coordinates), and how transversal through the screens of an app affect screen load time allow for tuning of apps to remove potential resource hogging.

It is especially effective to correlate the measurements of mobile device resources with the functional tasks within the mobile app. Being able to associate a spike in memory usage with a specific functional activity in the app is invaluable for developers to quickly pinpoint the area of the app code that needs to be addressed in order to resolve excessive resource usage.

User Sentiment as a Measure of Quality

Once your app has been released into production, your end users will begin to postreviews in the app store. App store review comments offer a rich source of quality assessment also.

If your mobile app has only a couple dozen reviews in the app store, it is easy enough to read each one and gain insight into the sentiment of your users. But once the number of reviews reaches several dozen or more, you need an analysis tool to effectively and efficiently mine the key insights from that amount of data. This app store review user sentiment analysis can shed light on trends in the perception of your mobile app audience, and can be useful to correlate with the other quality assessment measurements such as crash data captured from production versions of the app.

For example, the IBM MQA service includes an app store review analysis capability that captures all of the review text and searches for a set of special "user sentiment" key words in each review text. Analysis for app store reviews is organized into 10 distinct "attributes" about your mobile app, such as:

- Usability
- Stability
- Performance
- Elegance

You can drill down into each of the user sentiment attributes for your mobile app and see the analysis used to produce the score for that attribute, even going so far as to see the specific list of reviews containing comments about that attribute.

This app store user sentiment is invaluable, especially when you correlate the sentiment expressed by the reviewers with hard technical evidence in the crash reports and in-app user feedback records.

Some users would not take the time to post a review in the app store, but will comment about their perception of your mobile app in their various social media networks. Other sources of user sentiment include comments made by your users in social media such as Facebook, Twitter, or LinkedIn. Social media analytics can be used to uncover information about what your users are saying to their community about your mobile app.

Summary

Our point-of-view for a comprehensive solution for mobile testing and quality management is an approach that encompasses the full spectrum of mobile test techniques currently in use in the industry. The key elements of this solution are:

Run a suite of automated mobile tests against each build of your mobile app.

1. After sufficient automated testing has completed successfully, distribute the app to a group of human testers to perform interactive manual testing on the app.

2. Organize and manage the various test execution tools for mobile app testing (both mobile frontend and supporting middleware and backend services) using a test management product such as IBM Rational Quality Manager. Consolidate and link the test results from these multiple execution tools into a single mobile app quality metric, and link data from test case failures back to development work items for defect removal.

3. Use service virtualization, such as available in IBM Rational Test Workbench, to isolate various tiers of the mobile app so that testing can be concentrated solely on those specific tiers. Test the code on the mobile device without needing to have a complete middle-tier up and running. Test the middle-tier mobile app logic without having to coordinate large numbers of mobile device clients.

4. Automate the tests for the code that executes directly on the mobile device using the IBM Rational Test Workbench automated mobile testing capability (either directly or using IBM MobileFirst Platform Studio development environment). There is no need to hire specialized skilled programmers in order to produce the automated versions of your mobile test cases.

5. Concentrate your investment in real physical devices to only the highest priority device types and OS release levels. Rent the other permutations from a device cloud vendor that is integrated with your test management solution (such as IBM Rational Quality Manager) so that the same consistent tests can be applied to the "cloud devices" as is used for your in-house physical devices.

6. Use emulators as target test devices for your every day automated regression testing. Use the same automated mobile test capability to execute emulator test cases as you use for the real physical devices. An automated mobile app testing solution, such as IBM Rational Test Workbench, should work on both emulators and real devices.

7. Organize your manual test cases into logical suites and reduce the costs and increase the efficiency of your manual test efforts using IBM Rational Quality Manager manual test management capabilities.

8. Instrument your mobile app with IBM MQA code so that your manual interactive testers can submit bug reports directly from within the app being tested on their mobile devices.

9. Employ IBM MQA mobile app crash data capture services to obtain deep technical context information about each crash that occurs in your mobile app, whether during preproduction testing or after the mobile app is released into production use.

10. Leverage "in-app" user feedback to make it easy and efficient for your end users to communicate with your development team about their perception of the mobile app. Capture the context of the app and the device on which it is running, every time a user submits feedback.

11. Use an analysis tool to aggregate and gain insight from the app store review comments for your app. Correlate this app store user sentiment analysis with the other quality assessment data that you continue to gather about your mobile app in the field.

CHAPTER 7

Best Practices of Mobile DevOps

The pace of development and release for mobile apps is more rapid than previous forms of software and therefore your team must employ practices that wring every bit of inefficiency out of the develop-and-release lifecycle. Those practices that accelerate your team's delivery are generally called "DevOps." This chapter goes into detail covering the topic of DevOps best practices, and specifically how to apply them for mobile app projects.

What Is DevOps?

DevOps is an enterprise capability for continuous software delivery that enables organizations to seize market opportunities and reduce time to customer feedback. DevOps provides an approach for continuous delivery of software-driven innovation.

By applying lean and agile principles across the software delivery lifecycle, DevOps helps organizations deliver a differentiated and engaging customer experience, achieve quicker time to value, and gain increased capacity to innovate (Figure 7.1).

It is important to remember that the DevOps principles span the entire software delivery lifecycle. The spirit of DevOps is expanded collaboration among all stakeholders, not only between development and operations, but also among lines of business, suppliers involved in software delivery, and consumers themselves. In other words, the DevOps lifecycle is the software supply chain. All tools, processes, and stakeholders fall under its umbrella.

Some Definitions

The list of terms defined below will be used in this chapter:

- **Continuous Integration**—The practice in which software developers continuously, or frequently, integrate their work with that of other members of the development team. Continuous integration does not imply that every "change" to code is releasable or even deployable, but rather that the change did not cause an unforeseen breakage of functionality. The extent to which this is proven is determined by how much validation is built

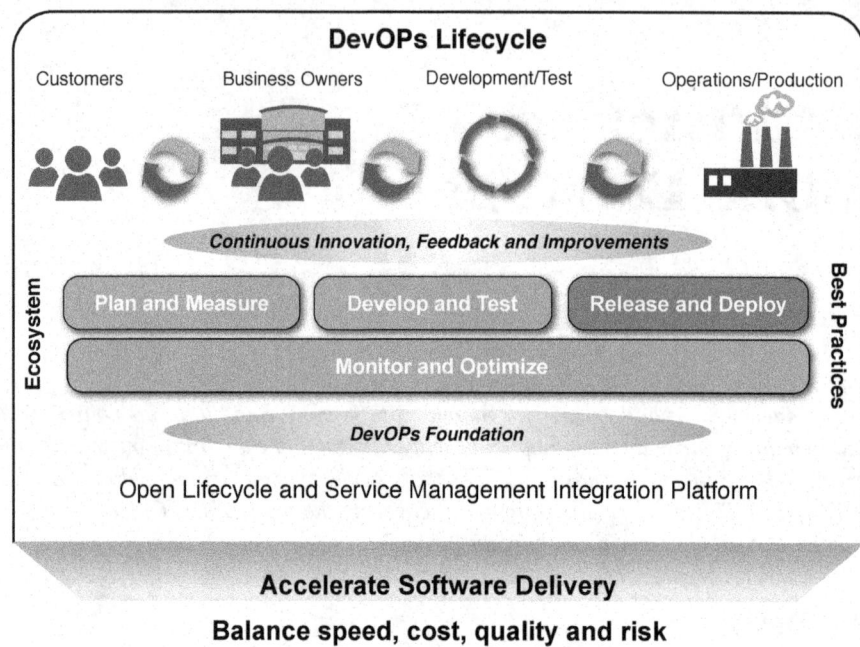

Figure 7.1   An overview of the DevOps lifecycle

into the continuous integration process. The fact that code "compiles" is one measure of success. Successful passing of unit tests, or even better, some type of integration smoke tests, is a better indicator that the change did not cause a break.

- **Continuous Delivery**—The practice of automating the deployment of software to integration, testing, staging, production, etc. environments. Automation provides a repeatable machine-driven set of steps that reduces the tremendous amount of risk involved when humans must execute deployment processes. In addition to the removal of risk, continuous delivery provides a much faster deployment process in a time when speed-to-market continues to be a significant business driver.

- **Continuous Testing**—The practice of inserting testing into the continuous integration and delivery automated processes. The automated testing inserted is dependent on the type of validation desired. For example, unit testing as part of an automated build process (continuous integration) can give instant feedback on the usability of the build. Only builds that pass unit testing are valid for moving to the next step in the pipeline. Another example would be the execution of a series of smoke tests after an application has been deployed (continuous deployment). Only applications that can be initialized and that pass a series of basic functional tests are worthy of more detailed preproduction testing.

- **Continuous Monitoring**—The practice of applying monitoring to servers, services, data, hardware, software, and users in all environments of the deployment pipeline. If benchmarks are established during QA testing, for example, differences can be noticed earlier and performance tweaks applied before reaching production.

- **Build and Deployment Pipeline**—The practice of establishing (and automating as much as possible) the steps and validation gates required to push a version of an application through its necessary environments, ultimately producing a production release.

- **Release Planning**—A critical business function that outlines the release roadmaps, project plans, and delivery schedules as well as end-to-end traceability across these processes.

The IBM DevOps Solution

Driving innovation throughout your business takes more than new technology tweaks. Real innovation demands a different approach; a software delivery lifecycle integrated into the culture of your organization that unites trusted, dependable, and secure multi-platform applications with continuous software delivery. Enterprises that master advanced practices of software delivery achieve better business performance than those that do not. Traditional software delivery approaches that address only a single phase of the lifecycle are not sufficient. They simply shift the bottlenecks creating unnecessary waste and delays. The IBM DevOps solution addresses culture, process, and tools integration across the software delivery lifecycle from ideation to delivery. This bridges the divide among key stakeholders to establish a powerful and essential enterprise capability for continuous software delivery that leverages lean and agile principles removing waste, reducing time to customer feedback, and accelerating software delivery.

DevOps is even more important in the context of enterprise mobile applications. Mobile applications are sometimes the one and only system of interaction a customer desires. A relatively short but bad user experience can lose that customer for good.

The IBM DevOps solution can be broken down into the following topics:

Plan and Measure

All too often, the minutia of the software delivery lifecycle takes on a life of its own and before long the business value of a long running project begins to be questioned. Soon the point of no return has been reached and the project lives on to save face. To combat this common experience, organizations should establish a culture that strives to constantly reduce waste and increase business value by managing the end-to-end lifecycle. This can be achieved by continuously planning, measuring, and bringing business strategy and customer feedback into the development lifecycle. As the marketplace works today and with the business climate we face, organizations need to insure that they "do the right things" and focus on activities where they will gain the most value, and strive to "do things right" within the boundaries of organizational strategy and business objectives. Employing lean principles will help teams start small, identifying the outcomes and resources needed to test the business vision and value adapting and adjusting with agility.

Develop and Test

Keeping a competitive advantage requires the continuous innovation of ideas and the ability to translate them into products and services that bring value to your customers. Collaborative development and continuous testing support the evolution of a business idea into a high-quality software solution by applying lean principles, facilitating collaboration among all stakeholders, and striking the optimal balance between quality and time to market.

By establishing a solid development and test strategy, companies can ensure that their teams stay productive, remove unnecessary project risk, and slash the cost of software delivery.

Release and Deploy

The bread and butter of DevOps is the automation of the release and deployment processes. Tremendous business value can be achieved by automating these processes that drive down costs by eliminating errors and rework, speed time to market by the simple fact that machines can perform repeatable deployment tasks faster than humans, and reduce risk by delivering higher quality application releases with increased compliance and end-to-end transparency.

The hard costs of software release and deployment can be easily measured and applying a DevOps strategy to this aspect of the software delivery lifecycle provides a quantifiable business value return-on-investment.

Monitor and Optimize

The customer feedback loop is a key component in insuring that the right solution is delivered to the customer. Improved visibility and continuous feedback across the development lifecycle will help you exceed service levels. Monitoring solutions offer enterprise-class, easy-to-use reporting that helps developers and testers understand the performance and availability of their application while optimizing capabilities. The goal is to maximize the value of every customer visit and ensure that more transactions are completed successfully.

DevOps Best Practices

As has been already mentioned, DevOps is a series of principles applied across the software delivery lifecycle. The following is a list of prescriptions that can be applied universally and that all contribute back to the goals of DevOps:

Plan and Track Everything

The planning aspect of agile development is well-documented. A maniacal focus on planning is a direct attack against risk. DevOps brings to the table tools and techniques that promote faster processes (automation) and quicker customer feedback. Therefore, it is even more important that you have at your fingertips the information needed to make quick course corrections. Following DevOps principles will bring information back to the decision makers quickly. Therefore, the decision makers need to be prepared to react. The more data you have on the status of everything (see the next section) will allow those direction changes to occur with confidence.

But it takes effort and tooling to track things. It needs to be easy for a tester to submit a defect so that it gets into the backlog of things to fix. It needs to be easy for an analyst to inject new requirements into the process so they get evaluated, prioritized, and available for the next sprint. Planning and tracking takes effort that sometimes requires a dip in productivity until the benefits are realized. However, having a long range attitude towards planning and tracking will pay off.

Dashboard Everything

One of the tenets of DevOps is faster, or even more desirable, instant feedback. Faster and more intelligent decisions are required in order to seize opportunities and reduce time to customer feedback. Much of the decision-making data required to move forward faster is already present in numerous development tools, app store ratings, systems management tools, etc. Presenting this information at the fingertips of the decision makers is important.

Dashboarding data can be a difficult task. For example, if a release has made it to Pre Prod Staging, what type of information is required to determine if the release is ready for production? The obvious answer is the test results of the test cases that have been run on the release. But there may be more information needed, such as the features that are implemented in the release. Even though it may pass the necessary tests, it may not have the required functionality. This requires access to data from much earlier in the lifecycle which is typically not available on a deployment dashboard. Having access to all relevant data for a release (requirements, defects, test results, deployment environment readiness, etc.) is required.

Customer feedback is also very important. Mobile app stores today have easy to use rating systems for applications. Mobile app users are very accustomed to rating the apps they use after a very short user experience. A dashboard that provides a live look at an application's ratings is a key deciding factor in how fast you rollout your next release. A bad customer experience that produces a bad rating is cause to accelerate that next release to recapture the customer base. Having this data as part of a dashboard is critical to mobile application success.

Version Everything

This prescription should be taken literally. Things that are versioned can participate in a repeatable and automated process in a way that ensures the integrity of the outcome. Version control systems are made to handle the authorization and authentication of users and the versioned content that they store. This ensures that if we repeat the same automated process, we should get the same results if the content provided to the automation has not changed. Versioning everything also allows anyone, not just those in the know, to be able to recreate an outcome.

Mostly everyone is familiar with version controlling code. But there are many other elements that participate in the software delivery lifecycle that should not be left uncontrolled. This list should include software configuration, test scripts, data definitions, data loads, and data change scripts, build and deployment scripts, and even system configurations. In this day and age of cloud solutions, the system patterns that are used to create environments should also be versioned.

Automate Everything

Automation is about tools, and it provides the engine that allows the benefits of DevOps to be realized. All of the cloud computing, continuous integration, and continuous delivery value comes from automating manual processes to increase speed and insure consistency. Organizations should continually strive to increase their automation.

Automation is in various stages of maturity across the software delivery lifecycle. Test automation has been around for years and has been a holy grail for testing organizations. More too often forgotten in the testing arena is the need for test data automation as well. Test automation only works if you can insure that the test data needed for the test cases is automatically set (or reset). Build and deployment automation has been in the hands of script writers forever. There are now enterprise-class continuous integration and continuous deployment solutions that finally provide the environment for the reusability and scalability of scripts that work. Cloud solutions are beginning to provide infrastructure automation. The ability for a multi-node complex environment to be spun up quickly and correctly in a matter of minutes or hours is now a reality.

Combining all of this automation provides a new reality—the ability to provide a new complex environment running a complex multi-tiered application in a matter of hours in what used to take weeks or months. Automation provides a unique ability to react to change faster than ever before.

Test Everything

The more automation that you create, the greater the need for testing. Testing provides the information needed to pass through gates in your delivery processes. A dashboard should not only tell if a build was successful, but also if that build is functional by observing the results of smoke tests. A successful deployment should not end at the "application is deployed," but instead after the verification that the application passes some basic functionality tests. An environment in a cloud infrastructure should not only just be provisioned, but some type of testing should be provided to insure that the infrastructure is ready for deployment.

The testing community for years has done a good job of defining test categories (functional, integration, performance/load, user interface standards compliance, etc.) and there are numerous test automation solutions out there to reduce the strain on testing resources. DevOps strives to get smart about testing.

Monitor Everything

Monitoring software systems and servers is nothing new to the operations community. However, if the only focus is on production systems, you miss the ability to capture valuable information earlier in the lifecycle. Early detection of performance degradation may prevent an embarrassing rollout in production when your customers get to have their "first impression."

Reducing time to customer feedback can also be expressed in terms of operations feedback. If the operations team can observe a server's throughput statistics during a performance test run, it may provide valuable feedback that can stabilize a production deployment. This is even more important to mobile applications. Understanding bottleneck points in a mobile application during user acceptance testing is much more valuable than learning with real customers. Monitoring in preproduction environments is the key. Building in monitoring capabilities from the start will allow monitoring to take place at any level of testing.

Mobile DevOps Challenges

Enterprise mobile application development presents some unique challenges to implementing a DevOps strategy. The following is a list of mobile specific DevOps challenges:

Fragmented Platforms

Mobile application fragmentation can refer to the ever changing landscape of mobile platforms to target as well as supporting the ever growing versions of a single platform (such as Apple iOS or Google Android platforms). Fragmentation can also refer to differences in device platforms (screen sizes as an example), different programming environments, and even different app stores. Either way you look at it, fragmentation presents a challenge to the mobile application development space. Since this is a chapter on DevOps, we will not discuss the pros and cons of targeting various platforms or how many versions of Android to support, but instead address the general problem of supporting more than one target infrastructure.

Finding solutions to this problem can begin with your development environment. The silver bullet of "build once run everywhere" does not look like it will be here anytime soon. However, there are solutions that can get you closer. IBM Worklight provides a development environment that provides mechanisms to separate common code from platform-specific needs. At a minimum this prevents your need for duplicate code. However, you still must be conscious of what is platform specific and what is not.

Automation can be a large part of a solution for fragmentation. Build automation insures that you are pulling platform-specific content from the correct location when building deployment packages. If left to a human, the potential for error is magnitudes greater. Deployment automation insures that the correct application packaged for a particular app store gets to that app store. Again, the more left to the human the bigger the potential for error.

Mobile Applications Front a Complex Enterprise Back Office

The introduction of the concept of systems of engagement (SOE) by Geoffrey Moore[1] has helped to highlight the shift in thinking of enterprise systems. System of Record (SOR) have been the focus of IT shops for years. Basically, SOE refers to the transition from current enterprise systems designed around discrete pieces of information ("records") to systems which are more decentralized, incorporate technologies which encourage peer interactions, and which often leverage cloud technologies to provide the capabilities to enable those interactions. Mobile applications definitely fall into the category of a system of engagement. Mobile applications are decentralized and encourage peer interactions to say the least (Figure 7.2).

---

[1] http://www.aiim.org/~/media/Files/AIIM%20White%20Papers/Systems-of-Engagement-Future-of-Enterprise-IT.ashx

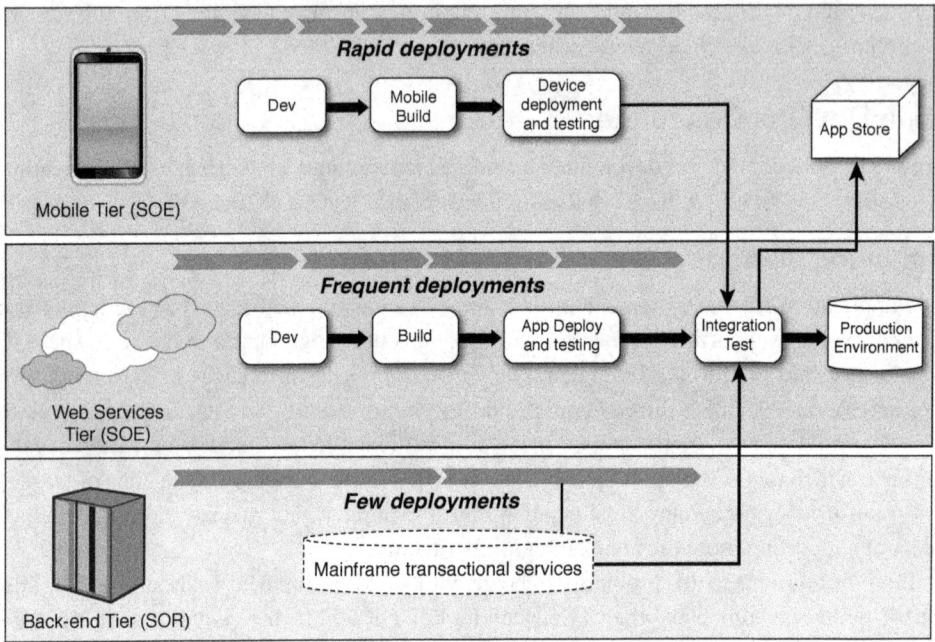

Figure 7.2   Multi-speed deployments for multi-tiered applications

The cadence of mobile application releases presents a challenge to up-to-now standard IT practices. Customers expect their mobile applications to be regularly updated providing new features and fixing bugs. Corporate backend systems, on the other hand, require lots of processes and approvals and are typically not updated that often. This presents numerous challenges.

- **Availability of back office systems for frequent testing**—The frequency of mobile application releases forces the enterprise to make available the necessary back office systems that interconnect with the mobile application. This isn't always feasible. This then creates the need for back office service simulation. The ability to connect to a simulated service to perform basic functional tests makes the frequent mobile release cycle possible. It eliminates the need for valuable human and technical resources and allows the necessary mobile application testing to occur.

- **Necessary updates**—The introduction of mobile functionality will often cause the need for an update to backend functionality. However, as was mentioned above, backend systems are typically not updated frequently enough to support the ever changing mobile capabilities provided by the mobile application. This forces the need for solving these problems in not the most idealistic architectural way and sometimes forces a "creative solution" in the mobile layer that should be made in the back office system.

App Stores Add Additional Asynchronous Deployment Step

Continuous deployment prescribes that you automate the deployment of applications to each environment. However, mobile applications typically have the app store sitting between a newly built application version and it being deployed to a device. This is an asynchronous step that has to be taken into account. This can even be more challenging in the iOS space as the time and process of having a new version of an application ready for use in the app store is out of your hands.

Security, Code Signing, and Keystores

Security is probably more of a concern for mobile applications than any class of applications before it. The speed of adoption and the lack of control present unique security challenges for mobile applications. And the risk is as great for the user as it is for the enterprise. Insuring that a user is installing the approved banking application instead of a malicious imposter is a legitimate public concern. Code signing and the like have been introduced as a way to alleviate this concern, but it presents a challenge to the DevOps processes. The process of validating certificates adds complexity to the testing process and must be accounted for.

Testing

Testing is always a challenge. However, great strides have been made in the IT space to automate testing and to interconnect the tester with the broader software development toolset. Mobile applications put an additional strain on the testing process. Mobile platform fragmentation and the growing list of supported devices have made browser compatibility testing seem trivial. Simulators and emulators also have a place in mobile testing but they typically fall short in both speed and functionality to the real device. Testing on the device is a must, but it becomes a challenge to capture application and device data when bugs are found. It is important to gather as much device data as you can to help developers fix the issue. Also, submitting bug information itself is a challenge when testing on the device. Testers are used to having their test tool at their fingertips to be able to submit bugs when they are found. If a bug occurs while the field tester is at home with a device, the ability to submit the bug becomes a challenge.

Mobile DevOps Best Practices

There are numerous best practices that have emerged when implementing a DevOps strategy. The following is a list that attempts to illustrate how they are applied to mobile applications and the challenges to overcome:

Practice Continuous Integration/Delivery and Automate Builds and Deployments

The mobile build and delivery pipeline might look like Figure 7.3. A few of the challenges listed above can be seen in this illustration. The different target platforms (and platform fragmentation) all must be taken into account during the build process. Deployments to different environments are also a unique challenge for mobile applications as different stages of testing require different

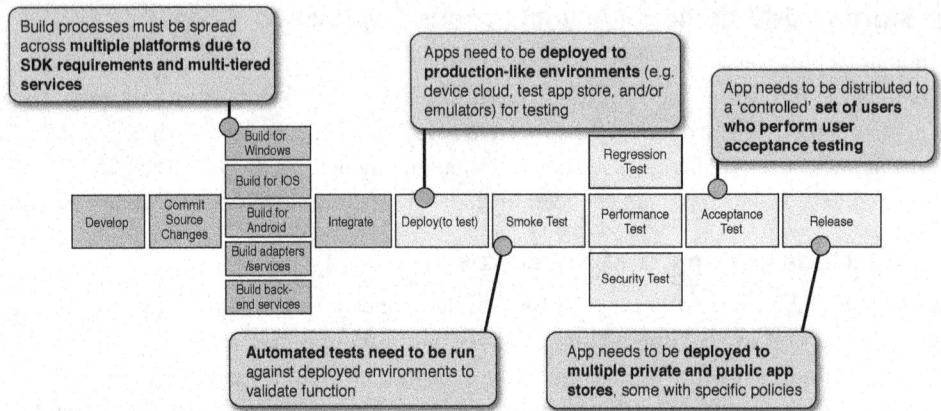

Figure 7.3   An example mobile app build and delivery pipeline of tasks

deployment targets. For example, early on for smoke testing, you might utilize an emulator and rely on a particular SDK for that deployment. In later environments, you need to deploy to a controlled set of users through an app store and the deployment process will be different. Release deployments might also include signing and/or the use of keys and the need to add some manual steps.

Mobile applications should require the same continuous integration rigor that you would apply to back office applications. Remembering that DevOps is all about faster feedback, continuous integration strives to insure developers get fast feedback on whether their changes have broken the build or caused tests to fail. By implementing continuous integration for mobile applications as well, these same benefits can be realized. The more automation you can include in your continuous integration strategy, the better (see below). Adding testing to your continuous integration process (JUnit, OCUnit, etc.) is even better. Taking advantage of automated testing solutions as part of continuous integration increases the value of the feedback.

It is safe to say that continuous integration is a valuable component to any mobile DevOps strategy. Continuous delivery, while more challenging for mobile applications, should also be included in your strategy. With the myriad of target platforms and variables that are no doubt part of each deployment process, an automated strategy for continuous delivery is critical. For example, deploying directly to an emulator or a device cloud may be OK at the early stages of testing, but using an actual app store (virtual or real) is most likely required when closer to release. The challenge of continuous delivery for mobile applications is, however, most likely worth the investment. Wasting time debugging deployment errors is a drastic hit to the fast and frequent delivery requirements of most mobile applications.

Test Each Build

Testing each build of the mobile app is an absolute requirement (Figure 7.4). Inserting as much testing and feedback into your continuous delivery processes is the key. However, testing mobile applications typically required numerous manual testers testing the application on different

Figure 7.4   An example of mobile app testing

emulators or devices. However, today there are automated mobile application testing solutions out there (IBM Rational Test Workbench, IBM AppScan, Appium, Robotium) that can be integrated into your continuous delivery process. For example, if a new mobile build could be pushed to a device and a series of automated functional tests could be run with the results collected during the deployment process, you could save yourself from wasting time on builds that may pass unit tests but fail on a category of devices.

Simulate Backend Services to Expand Testing Environment Availability

As mentioned above, a significant challenge to mobile DevOps is that most enterprise applications are not stand-alone applications that maybe only rely on phone capabilities, but are most likely a key "system of engagement" that relies heavily on existing corporate business processes. Enterprise mobile applications participate in and contribute to a significant backend workload on both business processes and enterprise data. Therefore, it is important to insure that the functionality is right. One of the tenets of DevOps is that your testing environments are as close to production-like as possible. In order to make this happen, you must have a strategy to "enable" as many services as possible for testing mobile applications. This can be challenging, especially for those services that have a cost associated with them such as subscription services or services hosted by a mainframe. Another challenge is that the service you are relying on may not be ready or available yet.

You can attack this problem by simulating backend services. Obviously you get your best value if your simulated service behaves as close to the real service as possible. As testing

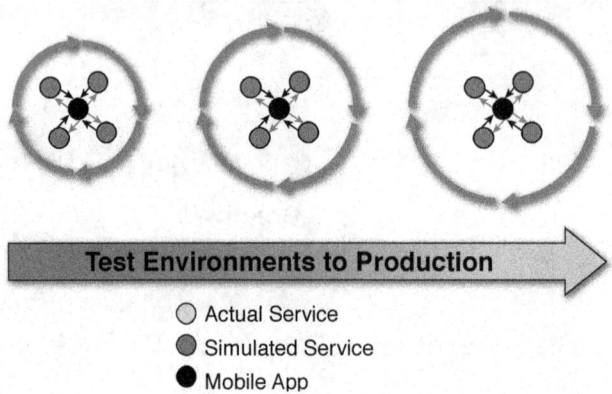

Figure 7.5   Replace simulated services with actual services on the path to production

environments are architected and designed, you determine what surrounding services are available. Typically it is difficult to provide all enterprise services for all levels of testing. Using service simulation allows you to provide more robust testing environments to meet the demands of the ever-changing mobile application (Figure 7.5).

From a DevOps perspective, it is also important that these simulated services are wired into the continuous delivery automation. This usually involves two details. First, you must wire the simulated service into the application as it is deployed to an environment. This may mean that configuration files are modified to alter service end points as a step in the delivery automation. The other detail involves insuring that the simulated service is up and running. This can also be added as part of the deployment automation (Figure 7.6).

Monitor Deployed Application and Backend Server Performance

Monitoring has traditionally been thought of as an operations problem. And therefore application performance issues are seldom caught until production or staging. The earlier you insert monitoring into your delivery lifecycle the sooner these types of problems can be captured (Figure 7.7). If server metrics are observed to be different during early QA testing, you have a much better chance of identifying and fixing the problem before it impacts customers.

Mobile applications impact performance in multiple areas. First of all a mobile application has an impact on the handset itself. A mobile application that performs poorly on a user's handset will no doubt get little use and be promptly removed. Thorough handset resource monitoring during handset testing can help to prevent a bad user experience. Gathering handset resource statistics during testing will help developers fully understand the impact of their code. Secondly, with the mobile application being the "system of engagement" for corporate business processes, they have a direct impact on backend servers, services, and data. Mobile applications should force the reevaluation of service-level agreements to insure that backend systems can handle the additional payloads that a global mobile user base will create. Monitoring during early testing stages will

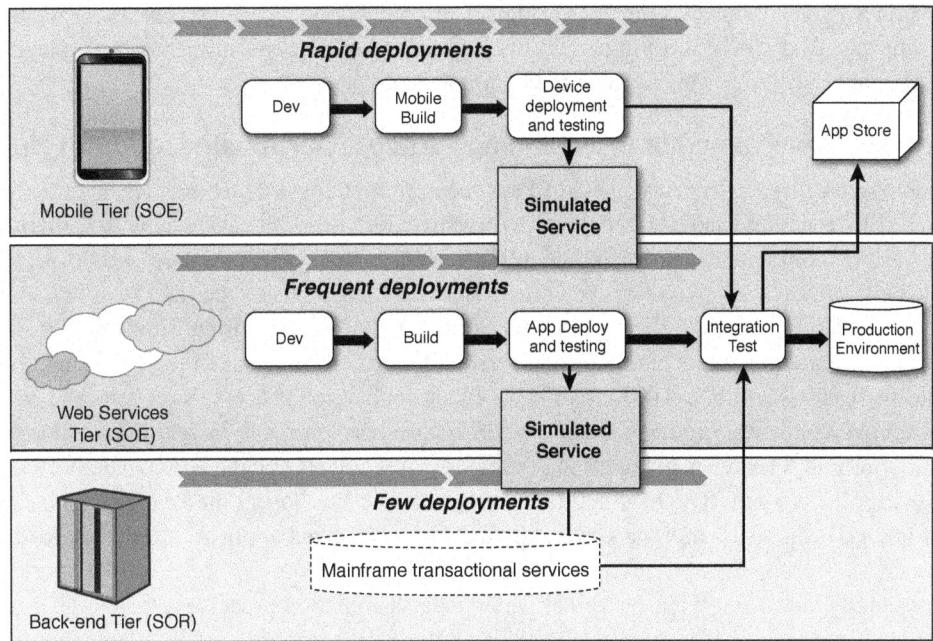

Figure 7.6   Where simulated services fit in the multi-speed delivery pipeline

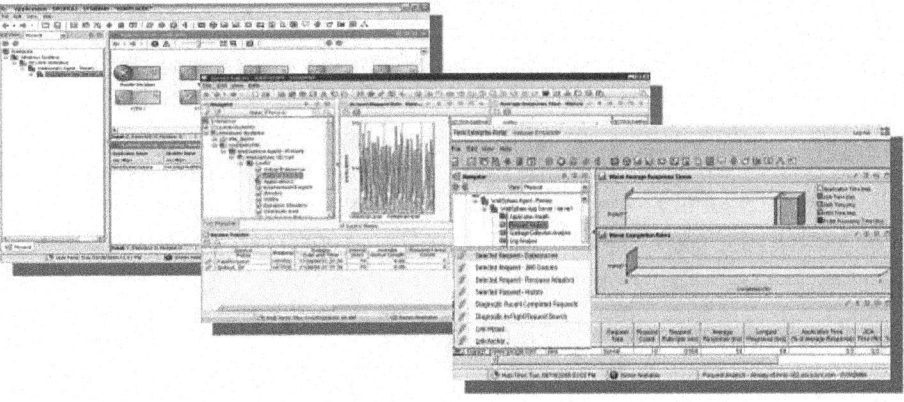

Figure 7.7   Example application monitoring solution

provide data to operations that they can compare to production systems to potentially identify future impacts. And maybe most important of all, monitoring of user sentiment can also help to alleviate a launch disaster. This is discussed in more detail below.

Centralize Governance of Provisioning Profiles, Certificates, and API Keys

Provisioning profiles, certificates, and API keys are in place to insure that the user or caller to a service, API, etc. is valid and not malicious. As organizations establish connectivity to app stores, devices, or APIs throughout the development lifecycle, different certificates may be used. As you move your application to production, you must insure that you are using the right keys.

The first step in getting this right is to establishing a centralized repository with appropriate security and governance to hold these valuable artifacts. For example, if your application needs to access a public API for part of its functionality, the process of registering for the use of the API typically produces a key or certificate that you must present during the API access at runtime. The key is given during the registration process by the API owner and along with it is typically some type of usage license agreement. By storing the key in a controlled repository, you can insure that the key is only used by authorized applications and authorized developers.

Secondly, insert into your build and deployment automation processes steps to fetch the correct keys and include them into applications and/or configuration files. By automating this process you can insure that the correct key is used at the right time by the right application.

Use a Private App Store to Test Deployment Devices

Deploying your mobile application to Apple's iTunes or Google Play is the final step in deploying your application to your ultimate end users if you are developing publicly consumable applications. This is not the time to test your deployment process to an app store. This process should be exercised and perfected in a private app store first. While this is not the forum to discuss the available private app store capabilities or contractual issues, it is important to stress the need for one. Again, one of the tenets of DevOps is to make each environment in your deployment pipeline as close to production as possible. And therefore insuring that you have practiced mobile application deployments to an app store is important for as many test environments as you have.

Convert Real User Feedback to Enhancements

Mobile applications that are distributed through an app store have the added benefit of capturing user sentiment as part of the app store experience. Mobile users are very accustomed to and expect the ability to provide feedback via rating the app in the app store. Development organizations have a unique and blatant feedback mechanism that needs to be taken advantage of. DevOps attempts to increase feedback loops to shorten delivery cycles and improve customer results. By translating real user feedback directly into enhancement requests (including the average rating or "number of stars") you will directly impact the delivery schedule and solidify the prioritization of changes.

Summary

Mobile applications are now a mainstay in most every industry and have had a dramatic effect on software development strategies as the term "mobile first" indicates. Taking a mobile first approach to development has put new a burden on traditional software practices necessitating the need for new thinking. Mobile applications are bringing the enterprise closer than ever to the customer and the demands of the customer are greater and more impactful than ever before.

DevOps has come along to provide the approaches and tools to help tackle this new challenge. By applying the mobile DevOps principles described above to process improvement efforts, organizations can reap the benefits of the fast moving mobile marketplace.

CHAPTER 8

Conclusions and Further Readings

Conclusions

In the previous seven chapters, we have progressed through a full cycle of development and release for an enterprise class mobile app. The reader will note that there are many considerations involved in the production of a mobile app that are just as relevant for an enterprise app as for a mobile app not related to any business organization at all. So, while the focus for this book has been on best practices within a corporate development environment, we believe that all of the concepts and concerns addressed in this book can be applied to any mobile application development project.

The process of software development, especially software as young and rapidly evolving as mobile apps, is constantly improving and changing. So many important aspects of mobile development changed during the time that we wrote this book, that we were challenged by the decision about when to declare it ready for publication. By the time you read the book, you may have learned about a new technology or solution that is relevant to the topic. The coauthors of this book continue to be involved in software, especially mobile software development, and we invite readers of this book to interact with us by posting comments on the "landing page" for this book on the IBM developerWorks community on the internet. The web address is www.ibm.com/developerworks/dwbooks/enterprise-mobile/index.html

While new tools and techniques for developing mobile apps will undoubtedly come (and go), there are some predictions that we can feel very confident about making. Mobile apps and the use of smart phones and other "smart" personal devices, in a wide variety of sizes and form factors, will continue to grow as the primary way by which the average person will obtain information and services over the Internet. There is no putting that genie back into the bottle. For businesses, mobile devices will be crucial touch points of engagement with their customers, likely to eclipse even the web browsers running on PCs and laptops.

But the apps running on these mobile end points will increasingly need connection and interaction with services and data available in "the Cloud." The Cloud, whether public or private, is the location where massive amounts of data can be collected, stored, and analyzed to uncover key insights. Consider image and video data as one example. While mobile devices will increasingly be the means for recording and viewing video, the physical limits of mobile devices will result in the cloud being the ultimate resting place for all of that information. In the cloud, the video data can be effectively combined and analyzed by multiple analysis engines in ways that are never going to be practical on the mobile device.

You can extend the video data analogy to all kinds of data: health monitoring data, geopositioning data, and data captured from intelligent end points of all kinds (appliances, household items, point-of-sale devices, etc). The cloud is the best, most economical location to store and process all of the data that will be produced by our universe of smarter devices. Yet when people want to look at the ultimate results produced by boiling down all of that Big Data, they will be looking at the display screens on their mobile devices.

So, even though mobile devices and apps will continue their inexorable pathway to become the main way that we interact with information technology, and will increasingly capture and collect a wider and wider variety of data, the true power and value of mobile software will only be realized when combined with cloud based services that can store the data and process it at scale. The implications of this assertion is that the mobile app development really means production of code for the mobile device and also, just as importantly, delivery of code that runs in the cloud. We have covered some cloud related material in this book, but just touched on current capabilities. Cloud computing is a domain of enterprise IT that is accelerating in its maturation. Perhaps, this would be a perfect topic for the third book in this series. Let us know by commenting on the landing page for this book!

Further Readings

The following list provides links to essential reference material related to each chapter of the book. Since web addresses can change over time, we recommend the reader to find the most up-to-date set of web links on the developerWorks landing page for this book, located at: www.ibm.com/developerworks/dwbooks/enterprise-mobile/index.html

Chapter 2: Development Lifecycle

1. IBM Design: http://www.ibm.com/design/

2. MobileFirst Platform: http://www-03.ibm.com/software/products/en/mobilefirstfoundation

3. Rational Application Developer wiki: https://www.ibm.com/developerworks/community/wikis/home?lang=en#!/wiki/2fad2df2-9c68-4aa3-abba-01e910211998

4. Rational Software Architect Design Manager: http://www-03.ibm.com/software/products/en/designmanager

5. Rational Software Architect: Model driven development and deployment of Node.js RESTFUL application https://youtu.be/fggfu8u_aOM

6. Rational Team Concert: https://jazz.net/products/rational-team-concert/

7. GrandSlam Tennis Reference Usage: http://www-03.ibm.com/software/businesscasestudies?synkey=A951466Y90458U66

8. IBM Mobile Quality Assurance: http://www.ibm.biz/mobilequalitycloud

9. Five Imperatives of Application Lifecycle Management: https://jazz.net/library/article/637

10. Development and Design Analytics: https://www.youtube.com/watch?v=NovQGpa2H6E

11. Offline Testing with a Mobile App: https://www.youtube.com/watch?v=Tm9MEk8dZzA

12. Mobile Testing: http://www-03.ibm.com/software/products/en/rtw

13. FURPS+: http://Wikipedia.org/wiki/Furps

14. DevOps Self Assessment: https://www.ibm.com/developerworks/community/blogs/invisiblethread/entry/diy_with_ibm_s_self_assessment_tool_for_devops_practices

Chapter 3: Design Related

1. IBM Design homepage: http://www.ibm.com/design/

2. *Standford Design Thinking*: http://dschool.stanford.edu/redesigningtheater/the-design-thinking-process/

3. Good Design is Good Business: http://www-03.ibm.com/ibm/history/ibm100/us/en/icons/gooddesign/

4. *The Design of Everyday Things*: Revised and expanded edition by Don Norman (http://www.amazon.com/The-Design-Everyday-Things-Expanded/dp/0465050654).

5. *Observing the User Experience, Second Edition: A Practitioner's Guide to User Research* by Elizabeth Goodman, Mike Kuniavsky, Andrea Moed (http://www.amazon.com/Observing-User-Experience-Second-Edition/dp/0123848695).

6. *Information Architecture for the World Wide Web*, 3rd edition by Peter Morville, Louis Rosenfeld (http://shop.oreilly.com/product/9780596527341.do).

Chapter 4: Mobile Development

1. DeveloperWorks Mobile Frontier blog: https://www.ibm.com/developerworks/mydeveloperworks/blogs/mobileblog

2. OMG Cloud Statndards Customer Council, Architecture for Mobile paper: http://www.cloud-council.org/CSCC-Customer-Cloud-Architecture-for-Mobile.pdf

3. IBM Bluemix Platform: https://console.ng.bluemix.net/home/

4. Developing applications for Android using Rational Team Concert, in an agile way: https://jazz.net/library/article/505

5. IBM MobileFirst: http://www.ibm.com/mobilefirst/us/en/

6. MobileFirst Platform homepage: http://www-03.ibm.com/software/products/en/mobilefirstplatform

7. MobileFirst Platform Developers: https://developer.ibm.com/mobilefirstplatform

8. AngularJS home: https://angularjs.org/

9. Apple Developer: https://developer.apple.com

10. Android Developer: https://developer.android.com/index.html

Chapter 5: Mobile Server

1. Wikipedia's definition for "Mobile Device Management": https://en.wikipedia.org/wiki/Mobile_device_management

2. Wikipedia's definition for "Mobile Application Management": https://en.wikipedia.org/wiki/Mobile_application_management

3. Comparing Mobile Application Management with Mobile Device Management: http://www.networkworld.com/article/2189066/tech-primers/mobile-application-manage-ment--mam--has-put-mdm-in-its-place.html

4. Mobile security best practices: http://www.cio.com/article/2378779/mobile-security/7-enterprise-mobile-security-best-practices.html

5. Wikipedia's definition for "Application Integration": https://en.wikipedia.org/wiki/Enterprise_application_integration

6. Configuring DataPower XG45 with IBM MobileFirst Platform Foundation (formerly IBM Worklight): https://books.google.com/books?id=MXjJAgAAQBAJ&lpg=PA224&ots=jNFE5x7IHX&dq=datapower%20xg45%20mobile&pg=PA224#v=onepage&q&f=false

Chapter 6: Mobile Quality

1. Rational Test Workbench: http://www-03.ibm.com/software/products/en/rtw

2. Build a mobile app that isn't perfect: http://www.ibm.com/developerworks/library/mo-build-imperfect-mobile-app/

3. What is Mobile Quality Assurance? https://www.ibm.com/developerworks/community/blogs/mobilequalitybeta/entry/what_is_mobile_quality_assurance

4. Wikipedia description of mobile application testing and various tools available: https://en.wikipedia.org/wiki/Mobile_application_testing

5. Get that 5-star rating for your next app: https://www.ibm.com/developerworks/library/mo-mqa/

6. First Glance: Mobile Quality Assurance: http://asmarterplanet.com/mobile-enterprise/blog/2013/10/first-glance-ibm-mobile-quality-assurance-mobile-analytics-developers.html

7. Setup Mobile Quality Assurance with Swift and iOS 8: http://www.ibm.com/developer-works/cloud/library/cl-mqa-swift-app/

8. OTA testing with MQA: http://www.ibm.com/developerworks/library/mo-otadistribu-tion-mqa-app/index.html

9. Mobile testing with IBM Rational Test Workbench, A step-by-step guide: http://www.ibm.com/developerworks/rational/library/mobile-testing-rational-test-workbench/

10. RTW Documentation: http://www-01.ibm.com/support/knowledgecenter/SSBLQQ_8.7.0/com.ibm.rational.test.lt.rtw.nav.doc/topics/c_ovr_rtw.html

11. Mobile application testing with Rational Performance Tester: http://www.ibm.com/developerworks/rational/library/mobile-application-testing-rational-performance-tester/index.html

12. Resource monitoring of mobile apps during automation testing: http://www.ibm.com/developerworks/rational/library/performance-analysis-mobile-applications/index.html

13. Test Automation of Mobile Application using Rational Test Workbench (RTW): https://www.youtube.com/watch?v=pRsRyR0Vj4g

14. How to create robust mobile and web UI test scripts: https://www.youtube.com/watch?v=MMsAs34N6PI&list=PLZGO0qYNSD4VZEYSvNbpXqFnAqmsSj2gQ

15. Running test workbench tests from IBM UrbanCode Deploy (for Continuous automation testing): https://www.youtube.com/watch?v=sWzew3fzhQg&index=8&list=PLZGO0qYNSD4VZEYSvNbpXqFnAqmsSj2gQ

Chapter 7: Mobile DevOps

1. Techniques for rapid mobile solution development: https://www.ibm.com/developerworks/mobile/library/mo-rapid-development/index.html

2. DevOps for mobile development: http://www.ibm.com/developerworks/library/mo-mobile-devops/

3. The Developerworks DevOps Zone: https://www.ibm.com/developerworks/devops/

4. DevOps for mobile apps challenges and best practices: http://www.ibm.com/developerworks/library/mo-bestdevops-mobileapps/

5. Implement enterprise scale iOS continuous builds with UrbanCode Deploy: http://www.ibm.com/developerworks/library/mo-mobile-devops-implement-enterprise-scale-ios-continuous-builds-with-urbancode-deploy/index.html

6. DevOps for mobile app development: https://www.ibm.com/developerworks/community/blogs/extendibminnovate/entry/devops_for_mobile_app_development

7. Mobile DevOps – why should you care? https://www.ibm.com/developerworks/community/blogs/mobileblog/entry/mobile_devops_why_should_you_care1

8. Mobile and DevOps: continuous change, user demand and effective collaboration: https://www.ibm.com/developerworks/community/blogs/invisiblethread/entry/mobile_and_devops_continuous_change_user_demand_and_effective_collaboration

9. Mobile and DevOps: creating a working environment of continuous testing: https://
   www.ibm.com/developerworks/community/blogs/video-portal/entry/mobile_and_
   devops_creating_a_working_environment_of_continuous_testing

This is just an initial list of reference material related to the chapters of the book. The online
version of this reference list is constantly being updated and revised to offer verified links to the
latest technological articles. Please visit that web page often for the most updated version: www.
ibm.com/developerworks/dwbooks/enterprise-mobile/index.html

Index

Note: Page numbers followed by 'f' and 't' denote figures and tables, respectively.